HOW DOGS WORK

How Dogs Work

RAYMOND COPPINGER
MARK FEINSTEIN

Foreword by Gordon M. Burghardt

THE UNIVERSITY OF CHICAGO PRESS : CHICAGO AND LONDON

RAYMOND COPPINGER is professor emeritus of
biology at Hampshire College. His books include
*Dogs: A New Understanding of Canine Origin, Behavior, and
Evolution*, also published by the University of Chicago
Press. **MARK FEINSTEIN** is professor of cognitive
science at Hampshire College.

The University of Chicago Press, Chicago 60637
The University of Chicago Press, Ltd., London
© 2015 by The University of Chicago
Foreword © 2015 by Gordon M. Burghardt
All rights reserved. Published 2015.
Printed in the United States of America

24 23 22 21 20 19 18 17 16 15   1 2 3 4 5
ISBN-13: 978-0-226-12813-9 (cloth)
ISBN-13: 978-0-226-32270-4 (e-book)
DOI: 10.7208/chicago/9780226322704.001.0001

Library of Congress Cataloging-in-Publication Data
Coppinger, Raymond, author.
   How dogs work / Raymond Coppinger and
Mark Feinstein.
        pages cm
   Includes bibliographical references and index.
ISBN 978-0-226-12813-9 (cloth : alkaline paper) —
ISBN 978-0-226-32270-4 (ebook) 1. Dogs—Behavior.
2. Canis—Behavior. I. Feinstein, Mark H., author.
II. Title.
   QL737.C22C64 2015
   636.7—dc23
                              2015015809

♾ This paper meets the requirements of ANSI/NISO
Z39.48-1992 (Permanence of Paper).

IN MEMORY OF ERICH KLINGHAMMER,
THE FOUNDER OF WOLF PARK

CONTENTS

FOREWORD

It is a great pleasure for me to compose the foreword to this book by Ray Coppinger and Mark Feinstein devoted to the ethology of dogs. It is doubly a pleasure as the book is dedicated to both Wolf Park in Indiana and the late Dr. Erich Klinghammer, who founded it. Erich Klinghammer was one of my teachers, a member of my dissertation committee at the University of Chicago and, subsequently, a long-time friend. Although he began his career by carrying out some of the first studies on sexual imprinting in altricial bird species, he always loved dogs, especially German shepherds. It was his dog Gitta who, over fifty years ago, found a pregnant garter snake on his research "farm" in northern Indiana. Knowing my interest in snakes, Erich gave her to me; the babies born to this snake led to my first snake research, resulted in my dissertation topic, and were a key factor in my subsequent career in serpentine ethology. Both Erich and Gitta were acknowledged in the first publication (Burghardt 1966); that Gitta was a dog seemed irrelevant to mention. So dogs and Erich greatly influenced my career. When Erich developed serious allergies to birds, he changed his scholarly interest to the behavior of dogs and other canids, advocated for proper captive conditions for wolves, championed wolf conservation, and promoted ethology by translating important books from German.

Ray Coppinger, first author of this book, I have known for almost as long, and with great fondness, since his sense of humor broke the tension of perhaps the most embarrassing event in my life as a scientist. This occurred at an American Association for the Advancement of

Science/Animal Behavior Society meeting in Dallas in 1968 when he followed me in the same session, presenting his dissertation work on birds. So Ray is also a refugee from the avian world, who with students and colleagues has studied the detailed workings of dog behavior. His important, but underappreciated, theoretical writings (Coppinger and Smith 1989) greatly informed my own thinking on the evolution of animal behavior, especially play.

This book on the ethology of dog behavior rides the crest of a renewed scientific interest in dogs and their evolution, behavior, cognition, and domestication. The great apes, humans' closest relatives, have been iconic mirrors for human behavior going back to the nineteenth century, including the famous World War I–era "insight" experiments of Wolfgang Kohler, captive studies of Robert Yerkes in the 1920s, the pioneering field research of Jane Goodall in the 1960s, the language trained apes of the 1970s, and the cognitively and socially adept apes so widespread today in both popular books and the burgeoning field termed "comparative cognition." But dogs are pushing their way into the august pantheon of super smart animals—to the extent that ape researchers are switching some of their efforts to our canine cousins.

Actually, however, the history of dogs as useful entries into the mysteries and origins of human behavior is quite ancient. Charles Darwin, who loved dogs more than any other animal, discussed breed differences in behavior and their scientific importance in his famous chapter on instinct in the *Origin of Species* in 1859. In his later *Descent of Man and Selection in Relation to Sex* (Darwin 1871) and in his book on emotions (Darwin 1872), he used dogs as exemplars of his three principles of emotions as well as arguing that dogs have attributes we think most distinctive of humanity such as loyalty, love, jealousy, pride, shame, imagination, reason, abstraction, and rudiments of language. In the seminal book on mental evolution by Darwin's protégé, George John Romanes, dogs, but not other carnivores, were ranked with apes as most humanlike in their mental attainments (Romanes 1883).

Konrad Lorenz, a key founder of ethology, also had a lifelong love for dogs, resulting in the first popular ethological book on dog behavior and evolution, *Man Meets Dog* (Lorenz 1954). Here a more scientifically measured and biologically objective treatment of dog behavior was presented. Interestingly, it took the cognitive revolution in comparative psychology and ethology to bring dogs back to the forefront of research, accompanied by modern molecular genetics working out the relationships among the various breeds, to make it timely to look closely once again at dogs through an ethological lens. *How Dogs Work* is a true milestone and worthy successor to Lorenz. It gathers together much of the massive amount of new biological and ethological findings on dogs, wolves, and their relatives and is authoritatively written by canine behavior researchers with a long history of seminal contributions. Furthermore, the authors focus on particular breeds rather than the generic "dog," maintain an ethological approach throughout, and, like Lorenz, present provocative, not consensus, views on many important aspects of dog behavior. Thus, not only will dog owners and aficionados learn much from this wide-ranging book about both dogs and science but dog professionals and canine scientists will have some persistent beliefs challenged as well.

This last claim derives from the fact that current approaches to comparative animal cognition using ingenuous methods not only highlight the cognitive complexity and problem-solving abilities of apes, monkeys, dogs, and other species but also facilitate the entry of uncritical anthropomorphic thinking into the interpretation of behavior. Such anthropomorphism, often considered "sins" that only pet owners and nonscientists commit, can actually enter the work and vocabulary of professional scientists themselves. This happens due to competition among researchers to show that apes, dogs, and other animals have humanlike abilities based on a tendency to view other animals through our psychology, not theirs. Such competition has led to continued debates between the promoters of animals having human smarts and killjoys who try to keep researchers grounded on the most parsimonious, even if strained and unlikely, interpretations.

Enter this stimulating book, which negotiates between cognitive and behaviorist extremes and discusses and interprets the behavior of dogs with the rich conceptual tools of ethology and basic animal learning processes. Although focusing on the fascinating behavior of sled, guard, and herding dogs, as well as wolves, the modes of thinking about behavior illustrated here can and should be extended to many more canine breeds and relatives. This book—unabashedly adapting and applying central tenets and methods of comparative ethology to understanding dogs—actually sets forth new ways of looking at the behavior of any and all species, including our own. It takes seriously Lorenz's insight that behavior is as much a characteristic trait of animals' species as are their anatomy and physiology. The wonderful photographs accompanying the text enable readers to interpret the postures, expressions, and behavioral dynamics of dogs and wolves, greatly enhancing understanding and enjoyment of these animals, much as being able to identify trees in a forest enriches the experience of hiking in the woods.

The authors, provocatively viewing dogs and other animals as complex machines, firmly tie the study of specific types and sequences of behavior, as well as learning, development, emotion, and cognition, to an intimate understanding of body, brain, and the evolution and modification of instinctive mechanisms. Today, molecular genetics is providing details of evolutionary origins, and neuroscience is providing insight into how brain processes underlie behavior. Alone they are often disconnected topics, but through behavior, and especially with an ethological approach, they can be more effectively connected than through any other behavioral science.

Gordon M. Burghardt

REFERENCES

Burghardt, G. M. 1966. Stimulus control of the prey attack response in naive garter snakes. *Psychonomic Science* 4:37–38.
Coppinger, R. P., and C. K. Smith. 1989. A model for understanding the evolution of mammalian behavior. In *Current Mammalogy*, ed. H. Genoways, 2:335–74. New York: Plenum.
Darwin, C. 1859. *On the Origin of Species by Means of Natural Selection*. London: Murray.
———. 1871. *The Descent of Man and Selection in Relation to Sex*. London: Murray.
———. 1872. *The Expression of the Emotions in Man and Animals*. London: Murray.
Lorenz, K. 1954. *Man Meets Dog*. Translated by M. K. Wilson. London: Methuen.
Romanes, G. J. 1883. *Mental Life of Animals*. London: Kegan, Paul, Trench, Trübner, & Co.

1 WHAT ARE DOGS LIKE?

This book is about the behavior of animals, and in particular about how dogs and other canids (like wolves and coyotes) "make a living"—what a biological organism like the dog actually does, and how and why it does what it does. We want to understand the forces and mechanisms that enable a dog to "tick" as it moves and acts in the world: why Border collies chase after sheep but livestock-guarding dogs don't; why greyhounds make good racing dogs but dachshunds do not; why a newborn pup behaves differently from an adult dog.

For us as ethologists—scientists who systematically investigate the biological bases of behavior—the notion of an animal "ticking" away, something like a clockwork machine, isn't just a clever metaphor. A machine is a device that does work by converting energy into action. Like any machine, a dog's behavior results from the translation of energy into patterns of movement (and ultimately, in the case of biological organisms, into offspring). How it is built, the shape and organization of its parts, and how it acquires the energy it needs to work all determine what a machine will do and set limits on what it can do. In this book, we will ask you to think about dogs and other animals in something of the same way.

You may well immediately and fiercely object that a dog isn't just some kind of mechanical wind-up toy. Surely, plenty of us believe, dogs have personalities and desires that we would never ascribe to a machine. Indeed, it is probably true that dogs and other animals do have "minds" that are at least something like our own. This is an exciting perspective; it is much in vogue in the popular media and it's the focus of a great deal of research in a new field that has come to be known as cognitive ethology. We'll look more closely at some of this work in a later chapter—but we will not often appeal to cognitive explanations in this book. Our aim is to see how much we can understand about why and how animals behave from the standpoint of "traditional" ethology: by considering how the bodies of organisms are constructed and how the shape of that biological machinery determines the patterns of movement and activity that are so important in their lives.

Since the Darwinian revolution, virtually all biologists—and most thoughtful people—have understood that all life on earth is related in an evolutionary web stretching over billions of years. The myriad properties of "biological machines"—the cells, tissues, and body parts and the processes that wire them together—are the result of evolutionary forces that have shaped and reshaped the genetic mechanisms that ultimately are responsible for translating energy into purposeful activity. Darwin's great idea was that evolution—by means of natural selection favoring beneficial variations in form—results in adaptations that enable an animal to acquire energy by eating, to avoid hazards (like being eaten by others), and to reproduce themselves. The central insight of ethology is that an animal's behavior, just like the organic parts that make up the physical form of a biological machine, is itself an adaptive product of those evolutionary forces.

If animals are indeed like machines in some sense, however, it goes without saying that they aren't just simple mechanical devices. Brains, for instance, are surely a critical component of the biological machinery that leads to behavior in higher organisms like dogs (and us)—and the vertebrate brain is quite possibly one of the most complex objects on earth, if not in the entire universe. There are some four hundred billion stars in the Milky Way galaxy—and sixty trillion neural connections in a human brain. The brain of a dog is not quite so cosmically large, but it is nonetheless a formidably complex organ—and it is only one part of an intricately integrated biological machine. Without bones and guts, skin and musculature, eyes and ears, and other organ systems—all products of the animal's genes, honed by evolution—an animal like a dog can't adaptively convert energy into effective action. So behavior must be a consequence of the animal's whole shape, the complex totality of inter-acting mechanisms that are constructed by its genes.

That said, the notion that animals are just (something like) simple machines does have a long intellectual history. A few centuries ago, the philosopher René Descartes famously articulated the doctrine of dualism, arguing that "body" and "mind" are two different kinds of thing, neither reducible to the other. He saw human beings as possess-ing both properties. But nonhuman animals, Descartes insisted, were essentially just like cleverly constructed clockwork devices that possess only mechanical bodies and no "soul-stuff." We are not going to take on this ages-old and still-ongoing philosophical debate about the relation-ship between body and mind. But it is instructive to think about why Descartes was able to view animals through the metaphorical lens of a ticking machine like the works of a clock—and how that idea might help us in understanding behavior.

Clocks are machines that mark out time. Inventive humans have come up with a wonderful range of ways to do this. Sundials indicate the passage of time by a shadow that relates to the position of the sun in the sky. Candle clocks, water clocks, and hourglasses can do the job because they use up various materials at a predictable rate. The me-chanical clocks that appeared in the Middle Ages, and were much re-fined by Descartes's time, worked on a different principle. These devices transferred mechanical movement—the swinging of a pendulum or the deformation of a spring controlling an oscillating wheel—to moving gears. Moreover, early clockmakers and other tinkerers soon discov-ered that their intricate mechanisms could do much more than just tell time—they could also cause other complex movements of many sorts. Eighteenth-century inventors thus delighted in building remark-able automata, clockwork machines that were intended to be realis-tic models of people and animals that could simulate lovers kissing, soldiers firing guns, or dogs chasing their own tails. Constructed only with gears that turned other gears, wires that pulled pieces into place, and pendulums that swung them away again, their machines could be made to appear to act in eerily but wonderfully recognizable ways. The tradition—now enhanced by sophisticated digital computational devices—continues today with the animatronic robots you'll find at

FIG. 1 A WIND-UP MECHANICAL TOY DOG "TICKING AWAY."
DRAWING BY CAROL GOMEZ FEINSTEIN.

theme parks around the world. Over the centuries, countless audiences have marveled at these amazing self-operating machines: "How lifelike they are!" Some of these robots can be hard to distinguish, at least at first glance, from the real thing. But the fact is that even a very simple mechanical wind-up toy may have something of the behavioral flavor of a real organism (see, e.g., fig. 1).

Two things make these artificial machines and devices seem like they are (almost) alive. They have the overt physical form of animals (or people); and they move like them. In effect, we see in these mechanical toys, these automata, the fundamental properties of animal behavior—which we define as "the shape of an organism moving in space and time." This may seem to you like an overly simple definition. But we think it's the right way to characterize what a natural organism is doing when we say that it "behaves." With that definition in mind, it makes perfect sense to say that a machine—even one that is human-made—exhibits behavior.

How it behaves is a matter of how it is built and how its shape changes as the machine engages with the world. When a mechanical cuckoo clock tells time by the movement of gears driven by a wound spring, it is behaving by the same definition that characterizes the behavior of a biological organism. The shape of its gear mechanisms, the relation in space of one gear and its teeth with respect to another, changes over time as gears turn and engage others. In one gear-shape configuration, the clock will chime four times. A little figure in a Tyrolean hat may also pop out, circle, and bow four times because a pulley and lever have been engaged by the shape of the gears, by their position and movement in space. As time passes and the spring imparts energy to the gears, the shape of the clock's innards changes—and its time-signaling behavior changes as well.

The analogy between human-built "ticking machines" and real animals can be quite compelling. Like a cuckoo clock that only strikes precisely on the hour and only on the hour, many birds exhibit courtship behaviors in which the male makes stereotyped (ritualized) movements in order to attract a female—for instance, a specific and fixed number of head-bobbing motions—that they perform only at a precise time of the year. Wolves come into season only once a year, in early winter. In any given year and area, their reproductive activity is synchronized and all of their pups will be born at just about the same time in early spring after an average gestation period of sixty-three days. Just as a striking clock is quiet between the hour marks, the wolf's stereotyped courtship behavior is dormant between seasons.

We must emphasize again that real animals are—of course!—astonishingly more complicated and more subtly constructed than any cuckoo clock or cleverly designed mechanical toy dog. Actual biological organisms are made of quite different materials. They have remarkable parts that can sense and respond to aspects of the world beyond them (though these days artificial automata can be built with this kind of ability too). A clockwork dog gets its energy from a human hand winding a spring or perhaps from a battery; real dogs get their energy from food that people provide for them. Wild animals need to find their own energy sources somewhere out in the world in competition with countless others. Real animals also change over time: a chicken starts out as an egg. A newborn pup and an adult dog have very different shapes.

Artificial devices, however, don't remodel themselves into something new. And, perhaps most important, animals can replicate themselves. This is something no human-made machine can do (yet). Reproduction is an all-important part of the biological story; it's at the heart of what we mean by "life." It's also an essential element in evolution and plays a crucial role in animal behavior.

It goes without saying, we hope, that animals weren't designed by a clever inventor who put parts together simply and logically. Rather, the shapes of organisms and their capacities for movement are complex outcomes of the forces of natural selection and other evolutionary and developmental processes. These led to a vast variety of solutions to the challenges of life; how and why these solutions work at all is often perplexing. In contrast, understanding how a clock or an automaton functions is a relatively simple task: you can patiently disassemble them, identify their discrete parts and work out how they function together. Biological machines are much more opaque. "Taking an animal apart," whether anatomically, physiologically, or behaviorally, is a daunting enterprise. The problem is that it's often unclear what the components actually are, what they're for, and how they fit together. From cellular mechanisms to neural organization, biological systems can be infuriatingly complicated; figuring them out has been the life's work of many generations of many kinds of scientists.

GENES AND THE BEHAVIOR OF BIOLOGICAL MACHINES

One thing that we do now understand with certainty, however, is that genes, the heritable chemical instructions encoded in DNA, are essential elementary components of the machine. They underlie its fundamental ability to replicate. An animal's genetic information plays a key role in establishing the initial basic plan of the machine (often in close interaction with the environment in which the genes operate). Over a lifetime, and on a daily basis as well, the genes build and rebuild the organism, specifying the character and limits of its body shape and its ability to move at any given time.

From that point of view, how an animal behaves is necessarily—and always—shaped by the genes that govern its construction. A dog behaves like a dog because it has dog genes—because it is built like a dog and not like something else. In this sense, all behavior is genetic. How an animal is affected by the world around it, the degree to which its behavior might be modifiable by training and learning, even the ways in which it can represent and use information (its "mind," if you will)—all of these are fundamentally limited by species-specific genetic characteristics that are, at its heart, the real subject matter of ethology.

We do have to be very careful and precise, however, when we say that all behavior is genetic. The actual patterns of movement that an animal exhibits are never explicitly "written" directly in the language of DNA itself; they are not laid out as such in the molecular code of a single gene. What the genes do is no more (and no less) than to trigger cellular mechanisms that give rise to proteins, building the animal's body and regulating bodily processes that enable it to move and act in certain ways. In this sense, all behavior does indeed have—and must have—a genetic basis. But at the same time, paradoxically, it's also right to say that there are no genes for behavior. What we mean is that there is no single gene for mate selection, no one gene that by itself controls the intricate motor patterns of predation. There are only whole bodies (and brains), built by the totality of gene expression, whose form allows for particular kinds of behavior.

So when we observe that racing greyhounds tend to run faster than dachshunds, does this mean it is a genetic property of the breed? In an important sense, the answer is yes. It is a result of the fact that a greyhound has a genome that builds an animal with the size, bone structure, musculature, and nervous system of a greyhound, giving rise to a shape that supports running fast. However, slow dachshunds don't differ from greyhounds because they have different genes "for speed": it is because dachshund genes build a different body with a different capacity for movement.

Go to a dog track, moreover, and you'll see that some greyhounds are clearly faster than others. An individual dog that is ill-fed or poorly exercised is going to lag behind in a race even if its twin brother, with identical gene sequences, was the winner. Genes interacting with a variable environment can have a profound impact on the animal's shape and its subsequent behavior. That said, while a developing greyhound's shape may change in many ways in the course of life, it won't ever turn into something that looks like a dachshund. And no dachshund will ever run as fast as a greyhound, no matter how much you might try to enhance its development and fitness—or train it.

Ultimately what makes a greyhound tick like a greyhound, or a dachshund tick like a dachshund, is its genetically predetermined overall form. What ethologists want to understand is how that form, and the biological machinery it encompasses, arose in the course of evolution, how it might change over a lifetime, and how it enables the adaptive patterns of movement that constitute behavior.

WHAT DOESN'T MAKE DOGS TICK: A FEW CAUTIONARY TALES

Dogs are a great study animal for ethologists who want to formulate and test scientific hypotheses about the nature of behavior. They are ubiquitous, easy to observe, and often fun to work with. We think they also provide some telling examples of how not to think about the subject. Before we delve more closely into how to describe and explain behavior—what makes an animal tick—it will be helpful to step back and examine some popular ideas about dogs in particular (one might well call them myths) that can lead in the wrong direction.

MAN'S BEST FRIEND?

Lots of people are satisfied to sum up the character and behavior of dogs by invoking the old, very tired, and probably misleading aphorism that they are Man's Best Friend—that dogs have a strong special bond with human beings, and their very essence makes them steadfast and loyal. (Of course, it isn't just "dog people" who stake a claim to man's best friend: people who love horses doubtless believe that the horse deserves that title.)

This picture is incessantly reinforced in popular culture and in the mass media. We all know pet animals that fit the image of the dog as a cherished humanlike friend. Most of us recoil in horror at the thought of using our dog friends for food (though some cultures eat dog, and others think horse meat is great, too); instead we spend billions of dollars a year to keep them in dog chow. Lots of dogs seem to live so comfortably and happily with people that it's not surprising we can think of them as bosom buddies.

We submit, however, that this sentimental view of the dog's behavioral nature is deeply flawed, and it doesn't help us understand why dogs actually act the way they do. The fact is that the human-dog relationship is no bed of roses. There are countless problematic dogs that are anything but good friends to humans. A massive dog-training and pet-psychiatry industry has arisen to try to fix undesirable behaviors and personality traits, recruiting a raft of fashionable techniques and medications for behavior modification. Scores of books have appeared on this "new revolution" in training dogs. That hasn't helped the five million dogs a year that end up in animal shelters or are euthanized because they can't be trained or are deemed dangerous. Neither does it hint at the 17 percent of the pets that are treated by veterinarians for sometimes very serious behavioral problems. Indeed, dog bites have become a virtual epidemic—we're tempted to call it a worldwide pandemic. In the United States alone, the bite rate has risen to 536 bites an hour—around 4.7 million dog bites per year. Some eight hundred thousand of those people require medical attention, and six thousand need hospitalization: it is said that dog bites are the second largest public health problem in this country.

Moreover, while there are probably close to one billion dogs in the world, less than one-quarter of them are actually what the World Health Organization would call "dependent and restricted." More than 750 million other dogs make a living by scavenging on human feces and refuse in alleyways and garbage dumps—and the occasional corpse. These free-ranging dogs, living and reproducing independently of human owners (or "best friends") on city streets and in the outskirts of rural villages, are the primary cause of some seventy-five thousand human rabies deaths per year around the world. As we write, a new rabies epidemic has started in the Republic of the Congo. Man's best friend, indeed!

Why then does the sentimental fiction persist so strongly? Part of the answer is that humans have an astonishingly powerful inclination to make sense (or to believe we are making sense) of the world by applying human categories to almost everything. We are strongly disposed to think about nonhuman animals and their behaviors, natural events like storms, or even inanimate objects like automobiles as having human-like attributes. This is anthropomorphism, the propensity for giving things (as the word's Greek etymology suggests) a human shape. It's a persistent and powerful worldview that seems to be deeply rooted in our psyches. To a small child, a rag doll or a plastic action figure can seem just as real and humanlike as their human friends. Making them move and talk seems perfectly reasonable to children. Adults are not immune. Safely home after a long and harrowing car trip in a snow-storm, we pat the car on the hood and tell it, "Good job!" When computers or kitchen appliances fail, we curse them, kick them, and try to cajole them into behaving properly. In large measure this is why we can see clockwork automata as so lifelike.

If pressed, of course, we'll acknowledge that these kinds of ma-chines don't really respond to our actions. We know they don't really understand us (or really love us) even as we persist in treating them as if they do. There is, of course, little about a car or a computer that is really "like us," but it is very hard for us not to think that there might be. Some say this impulse arises from a deep need in our psyches to assign value and familiarity to things that are important to us. Whatever the expla-nation, anthropomorphism is prescientific mythological thinking—the sort that led early humans to try to understand volcanoes by inter-preting their eruption as the expression of an angry personified god.

It's especially easy to anthropomorphize living animals that are, in fact, like us in certain ways. Seeing a reflection of yourself in a tarantula or a sea slug may be a bit of a stretch. But it doesn't seem so far-fetched to impose "human shape" on a mammal like a dog with body parts (hair, four limbs, and nipples) and processes (birth, care of offspring) that closely resemble our own. It seems particularly easy to anthropo-morphize animals that live in especially close association with us. So we give our dogs (and cats and parakeets and horses—that other candidate for man's best friend) humanlike names, talk with them as if we're having real conversations, and knit them sweaters. We *want* to think of animals as having something like human thoughts and emotions. We *like* the idea that a dog might actually feel a "firm friendship," as Darwin himself once put it, for us or for other animals like sheep. We have friends, so dogs can have them too, and they must somehow feel the same way about us as we do about them. But we have to be exceedingly careful about thinking this way and using this kind of language about dogs or any other animal. Looking at nonhuman animals from a human perspective may sometimes help us to come up with interesting hypotheses (inspired guesses that still require careful scientific scrutiny), but it can often lead you to look at behavior in ways that may just be plain wrong.

A classic case in point is the iconic British tale of Greyfriars Bobby. Bobby was said to be a Skye terrier who belonged to a night watchman in mid-nineteenth-century Victorian Scotland.

The usual story is that after its owner died and was buried in the cemetery of Greyfriars Church in Edinburgh, Bobby faithfully sat by his grave for fourteen years. After its own death, statues were erected to the dog's fidelity and love for its master (fig. 2).

The story of Bobby has warmed the hearts of British dog lovers for a century and a half, and it spawned films, books, and a lively (and lucrative) tourist trade. There is mounting evidence, however, that this iconic sentimental tale of a boundlessly loyal animal, a friend even after death, is really a myth. Jan Bondeson, a historian at Cardiff University, has concluded (as reported in the London *Telegraph* newspaper in 2011), that the original Bobby was likely just "one of around 60 Victorian cemetery dogs who waited around for food in graveyards and were so well treated they that stayed there to lead an independent and comfortable life." Over the years, local merchants, in fact, recruited a series of different animals, resembling Bobby, to keep playing his role—and supporting tourism. Indeed, around the world free-living dogs, with no owners to "love," are often found near newly buried bodies (see plate 1). Perhaps cemetery visitors or caretakers feed them; not infrequently, they consume human remains.

FIG. 2 MONUMENTS TO GREYFRIARS BOBBY IN EDINBURGH. "MAN'S BEST FRIEND" IS GOOD FOR BUSINESS. PHOTO BY EVIE JOHNSTONE.

We also go astray if we buy into the notion that dogs are really just wolves that somehow have come to be able to live among us. Perhaps this idea taps into a modern human desire for a closer connection with nature—but it's just another kind of myth. It's true that when you look at dogs (or at least some of them), you can see what seems to be an obvious surface resemblance to wolves. And ask just about anyone about where dogs come from and you'll almost certainly get the answer, "They evolved from wolves, of course." Read almost any media story about the subject and you'll be told straight off that most scientists also believe the dog descended from the wolf. Wolves and dogs are certainly phylogenetically close relatives, along with coyotes, jackals, Ethiopian wolves, and dingoes; they can all interbreed and produce viable and fertile hybrid offspring. But there is actually little evidence that dogs are direct descendants of those iconic big gray wolves in northern Canada or Russia—at least, not from a population of animals that acted just like the wild wolves we see today.

More importantly, modern dogs simply don't behave like modern wolves. Popular dog "experts" like Cesar Millan may tell you that a good dog owner needs to play the role of the alpha wolf, the dominant pack leader. But the fact is that dogs don't live in hierarchically organized packs. Indeed it's doubtful that most wolves live this way. And dogs are different from wolves and other wild types in countless other regards. Adult dogs can be readily trained to respond to human commands and to carry out myriad jobs for us—but not wolves. Wolves can be great natural problem solvers; they, like coyotes, are consummate escape artists when penned up by humans. Dogs, in contrast, are easily caged. And consider how the several species play out their parental roles and care for offspring. Male and female wolves form pair-bonds, living and mating together while defending a feeding territory, often for life (and typically for fewer than three reproductive years). Dogs never do. The males of all the wild types provide food and protect their mates while the females are nursing pups. Not dogs. Wild type mothers regularly regurgitate food to their young offspring; so do the fathers (who also bring back food from the hunt to feed the young). Dogs don't regurgitate often enough for pups to survive being fed in that way ex-

FIG. 3 DOGS IN THE MEXICO CITY DUMP. A BIG DIFFERENCE BETWEEN DOGS AND MOST OTHER CANIDS IS THAT DOGS CAN EAT CALMLY IN THE PRESENCE OF HUMANS.

clusively. And dog fathers don't have anything at all to do with raising pups.

The simple fact is that domestic dogs and wolves are different animals, adapted to different environments, and cannot live (well) in the other's niche. Wolves are consummate predators, but it's rare to find a dog that can hunt down and kill a moose for food. Dogs will track deer or chase wild hogs as they "hunt" with humans for sport, but it's highly doubtful they could ever earn a living by hunting on their own. For their part, wolves rarely become tame enough to get by in the household dog's world of human hearth and home. They may occasionally live on their own in or near human habitation, but they tend not to be able to eat in the presence of people. Whereas even free-living dogs that are raised in garbage dumps or just outside town are able to dine in the company of humans.

Yes, there are some tamed wolves, coyotes, and jackals. In our experience, however, it takes a herculean effort to get a wolf to develop in a way that makes it possible for the animal to be a tractable companion of humans. It's hard *not* to tame a dog. Just put in a few hours a week with a puppy during its fifth, sixth, and seventh weeks of life (even while the mother is still nursing them)—that's enough to make them orient to and socialize with humans. They'll hang around underfoot for the rest of their lives.

Taming a wolf is something else. You have to hand-rear them—the wolf pups will have to be bottle-fed—and begin to socialize them starting at ten days, before their eyes are open. Never past three weeks. If you start that late, a wolf may be able to tolerate some degree of close contact with humans but it won't form a social bond. It may be less skittish but won't greet humans or solicit attention even from familiar people. Starting at ten days you need to stay with wolves twenty-four hours a day every day through the fourth to sixth week of life, or "until you can't sleep with them anymore because they bite you when they wake up," laments Kathryn Lord, a former student of ours, now a professional ethologist and an experienced wolf tamer. For many weeks after that, you'll spend all your waking time with them, some eighteen hours a day. Taming a wolf can take many thousands of hours of work to produce a result that happens early, easily, and naturally in the dog. The romantic in us may like the idea that our placid household pet is really just a step away from the wild. Like the desire to see dogs as man's best friend, however, this is also mythological thinking. We need a better, more systematic, and less sentimental way of understanding what dogs—and animals in general—are really like.

2 WHAT MAKES ETHOLOGISTS TICK?

Ethologists are fundamentally concerned with what animals actually do in nature. We spend our time observing and describing animals as they move around in the world "making a living"—finding food to eat, avoiding being eaten by predators, interacting with other animals in competition for resources or to find a mate, giving birth, and raising offspring. The scientific goal is to build a theory of how animals are able to do all this. It's possible, of course, to study animals in the laboratory, to try to understand the limits of what an organism can do (or might be able to be induced to do) under more controlled and artificial conditions. Ethologists, however, have a special affinity for looking at animals on the ground and on the animal's own terms.

Many millions of fascinating species live in the natural world outside of human influence or control, and many ethologists are particularly intrigued by the exploration of behavior in the wild. Pick up any journal in animal behavior or ethology and you'll find studies galore of gazelles grazing on the African savannah or howler monkeys shrieking in the trees of the Amazon rain forest. Needless to say, however, we human animals are an awfully big part of the natural world, and these days it's a rare animal species that exists in the primordial "wild." Indeed, wild species have been relentlessly pressed by human population expansion and economic activity; many of their populations have dwindled in number (some to extinction) and they have become harder and harder to investigate. That's one reason why our study animal is the mundane, ubiquitous, and reproductively successful domesticated

FIG. 4   THE TWO COPPINGER BOYS TALK ETHOLOGY WITH GRETA AND KONRAD LORENZ AT
THEIR HOME IN 1978. PHOTO BY LORNA COPPINGER.

dog, a species with which humans have coexisted closely for better than
eight thousand years. They may be as ordinary and familiar a critter as
they come, all one billion of them, and we may often see them through
an anthropomorphic and sentimental lens. But, in fact, dogs are a great
model for illuminating the general scientific principles that explain
what animals—and ethologists—do.

Konrad Lorenz (1903–89) was one of the founders of the modern
field of scientific ethology (fig. 4). A Nobel Prize winner, a painstaking
observer of many species, and a lively and literate writer about animals,
Lorenz spent much of his long career looking at and thinking like a
scientist about dog behavior. That didn't prevent him from indulging
in a little sentimentalizing about man's best friend. In the last line of
his classic book *Man Meets Dog*, Lorenz describes his own pet Alsatian
as "an immeasurable sum of love and fidelity." There aren't many modern dog lovers who would find fault with such a charming tribute, and
Lorenz the dog lover may well have thought he could gain some mea-

sure of scientific insight from the vantage point of his profound sense of connection with the many animals he kept at his home and farm.

Nevertheless, when Lorenz won the 1973 Nobel Prize in Physiology or Medicine for his studies of imprinting in greylag geese (along with Niko Tinbergen for his work on herring gulls, and Karl von Frisch who discovered the "dance language" of honeybees), it wasn't because the Nobel Committee bought the idea that feelings of affection for and empathy with animals would illuminate the puzzle of how and why animals act as they do. No, the prize—the first to be given for scientific work on animal behavior—was awarded for the fundamental ethological idea, in the words of the Nobel committee, that behavioral patterns "become explicable when interpreted as the result of natural selection, analogous with anatomical and physiological characteristics."

This idea is the beating heart of ethology. It means that behavior itself is a consequence of organic evolution—as much a character of the adaptive machinery of an animal as its limbs or liver. It also means that particular behaviors should be viewed in just the same way as physical features—as characteristic taxonomic traits that define biological classifications. The term "ethology" itself comes from the Greek and means "the study of character," aptly capturing this core notion of the unity of behavioral and physical traits.

Indeed, one consequence of the ethological perspective is that natural selection can be understood as acting simultaneously on behavior and structure. It's not too far off the mark to say that, for ethologists, what evolution really "cares about" is swimming or flying or running. Fins or wings or legs are only a means to an end. When the first marine animal lumbered or lurched onto dry land, its selective advantage was the behavioral fact that it could use some (already evolved) property of its shape and structure to move in a particular way.

It follows from this way of thinking that ethologists focus their inquiry on the observation, measurement, and explanation of inherited behavioral patterns. Sometimes these have been called innate behaviors, or instincts. Neither term is much in vogue these days. "Instinct," in particular, has a whiff of the Victorian about it. The early ethologists didn't shy away from using the term (Tinbergen wrote a lovely book on the subject titled *The Study of Instinct*), but in hindsight this was probably a mistake. Among other difficulties, the vocabulary tends to obscure the fundamental ethological insight that behavioral and physical characters are the same kind of thing. Both are inborn heritable traits; but terms like "instinctive" and "innate," at least in common parlance, are typically reserved only for behavior and aren't applied to the physical characteristics of an organism. So consider how utterly wrong it sounds to say that "humans have an instinct for five fingers."

The term "innate" is problematic as well. It is commonly understood to mean that a trait is present at birth and fixed in its character by a specific genetic program. It may be a bit less infelicitous to describe a physical trait as "innate" than to call it an instinct. However, just as there are no single genes for aggression, there are in fact no genes for five fingers either. Fingers are the product of an initial body plan that undergoes a cascade of physical and chemical reactions as an animal develops. Slight changes in that cascade, even simple differences in the physical environment such as heat or cold or changes in the chemical environment like an introduced medicine or alcohol, can change the shape and even the number of fingers. The reason the vast majority of us humans have five fingers is that when that cascade of reactions took place we lived in very similar fetal environments; our mothers had body temperatures of 98.6°F, were healthy, and didn't ingest medicines or other chemicals that influenced development. The unfolding of genetic expression under the right conditions is what gave us our five fingers. Seen in this light, the trait can't be said to be strictly "innate" or "in the genes." It is the outcome of a developmental process that was initiated by a set of genes. Moreover, critics of the terminology of "instinct" and "innateness" have argued that it might never be possible to conclude that a behavior is entirely inborn. At every moment of its life, and even before it is born (as our example of five fingers illustrates), an organism is always crucially interacting with the world around it.

Since both words are so vexed, we will usually substitute the general term "intrinsic" in place of "instinctive" and "innate." Using the word "intrinsic" in this way is attributable to the embryologists who study development and growth. In a later section of the book, we will show why it provides a better characterization of inherited properties, whether behavioral or physical. We will also adopt two other terms that differ from the usual ethological vocabulary: "accommodation" and "emergence."

By "accommodation" (another embryological concept), we mean that biological structures and behaviors—the products of genes—are shaped and reshaped by one another as well as by the environment in which they develop. "Emergence" is the way that simple processes and properties can interact to produce structures and behaviors with (often much) greater complexity than the sum of their parts; it is a notion of increasing importance in the study of biological, physical, and computational systems alike. In chapters 7 and 8, we will explore these ideas more fully.

Our vocabulary and our view of ethology, then, is something of a departure from the traditional emphasis on the instinctual. We see behaviors arising from three different (though interrelated) forces and processes, and these three kinds of behavior—intrinsic, accommodative, and emergent—all need to be considered when we study dogs or any other animal.

NATURE VERSUS NURTURE?

These terminological arguments and refinements may well remind you of the long-standing and still-contentious nature/nurture controversy. For many years scientists of all stripes, and the public at large, have argued often heatedly about the relative importance of an animal's genetic inheritance and physical form (nature) and its development in a highly variable environment (nurture). The popular view, especially as it relates to humans, has seesawed crazily. One minute the public imagination is captured by the notion that there are "genes for religion" or inborn dispositions to violence and warfare. Then, a reaction sets in, and ideas about innateness are excoriated as examples of rank "biological determinism" that blithely ignore the role of culture or the possibility of change in human behavior. The problem is that, all too often, the question is viewed as a dichotomy, a debate with only two sides where one or the other must be primarily or entirely responsible for behavior.

Not surprisingly, perhaps, biologically inclined ethologists are often seen as lining up on the side of nature. Like all biologists, we take extremely seriously the Darwinian premise that animal species are adapted to their habitats by evolutionary processes. "Nature" is of course at the heart of this adaptationist stance. To be sure, not every biologist is convinced that biological adaptation tells the entire story of life and behavior. Some recent evolutionary thinkers (e.g., Steven Jay Gould and Mary Jane West-Eberhard, among many others) have challenged the primacy of adaptation via natural selection as the driving force of evolution and see developmental processes—or even accidental contingencies—as playing major roles. Increasingly, biological history is viewed as a product of the interaction of both evolutionary and developmental forces (an approach endearingly called evo-devo). As we will see in chapters 7 and 8, some of our own ideas about developmental accommodation and emergence also raise questions about the power of the standard picture of "strong adaptationism."

From a different standpoint, other scientists, among them many psychologists, have concerned themselves solely with how behavior changes (how an animal "learns") in response to the contingencies of experience during an individual lifetime. It's not surprising that they are inclined toward the "nurture" side of things; some reject entirely the idea that any behavior can in any significant way be predetermined by the genes.

In the final analysis however it seems abundantly clear that, like "instinct" and "innateness," "nature" and "nurture" are really the wrong words. An animal's phenotype—the sum total of all of its physical and behavioral traits—is always and necessarily the result of a highly complex synergy between genes (and their products) and the growth and experience of the organism. The availability of food resources, the presence of other animals, and even random contingencies like the weather all can have a profound effect on how an individual animal will look or behave at any given moment.

We jokingly ask our students, for instance, "Do you think you were genetically programmed to have the face you do?" (We owe this great example to the evolutionary biologist and geneticist Richard Lewontin.) The students usually reply, "Of course, and that's why I look like my father [or mother or granduncle]." There's no doubt that people often resemble their parents or other ancestors, and it seems reasonable to think that this similarity must be due to "nature," to genetic inheritance. Well, in an important respect a face must be genetic—it is, after all, a collection of skin, bone, and cartilage tissue that was certainly built by an individual's particular genes. But genes can't be the whole story. Faces change over a lifetime, sometimes very dramatically. Which face was programmed in the genes? Was it the way you looked when you were four years old? How about when you were seventeen? Or seventy-five? It's probably fair to say that people don't actually have exactly the same face two days in a row. All of your many faces may tend to more or less look like "you," but is there a set of genes for every single face you see in the mirror over your lifetime? Of course not.

In fact, all organisms develop as they live out their lives—their shapes change constantly through an exquisite interplay between the genes and between genes and the animal's environment. A mammal or a bird begins as a fertilized egg; the embryo enlarges and differentiates. At birth, a neonatal mammal or bird has a very different shape compared to the mature adult of its species. Each of those stages of life—the egg, embryo, neonate, juvenile, and adult—is working with the same set of genes, but those same genes somehow conspire to produce quite different outcomes as the animal grows.

Not only does the neonate have a body that differs from the adult's, it also has specific behaviors that are intrinsic to its particular stage in life. A newborn mammal suckles on its mother's teat (and doesn't have to be taught to do so); an adult eats solid food in a way that is quite different from infant suckling. Throughout development, intrinsic properties of the organism interact with one another and with the environment to produce new structures and capabilities: accommodation and emergence play critical roles. The bottom line is that no animal's genome holds a full blueprint spelling out the precise details of all of the changes in form and behavior that it will exhibit in life. Seen in this light there really can't be a simple binary distinction between nature and nurture: they act together. But even that conciliatory slogan is a bit too simplistic. What biologists want to understand is how the phenotype (P) of an animal arises. The genotype (G) is its genetic "nature." The influence of the environment (E) is what we mean by "nurture." Is the phenotype just an additive function of genotype and environment, nature and nurture? Put algebraically, is it a matter of $P = G + E$? If we want to know the contribution of nature, can we solve the equation for G by subtracting E from both sides: $P - E = G$? That is, if we look at the phenotype and subtract the role of nurture, are we then left with the influence of nature?

This used to be the way experiments were controlled—take two animals and raise them in identical environments; any differences had to be genetic. But suppose we want to know which breed of cow produces the most milk. So we raise two breeds—say Holsteins and Jerseys—in the same barns and provide them with the same food. Their genotypes are different but the environmental conditions in which they live are exactly the same. Under these circumstances the Holsteins give more milk. Can we therefore simply subtract environment from the equation and conclude that a genetic breed difference—nature—makes Holstein better milkers? Unfortunately, no.

While it may be true that Holsteins give more milk than Jerseys in a particular environment, if we change the conditions of nurture—say, we put both animals onto pasture to eat grass for several hours a day—Jerseys may outperform Holsteins. Again we have controlled for environment; but this time the Jersey gave more milk. So which breed of cow gives the most milk? Maybe the answer lies in another question: to what environment are they adapted?

The environments in which animals live are extraordinarily diverse and in a state of constant flux. Higher light levels could make a difference in milk production; temperature might have been a factor. Just as we've seen that analyzing the genome of an animal doesn't—indeed cannot—provide a complete account of all the details of shape and behavior that it will exhibit in its lifetime, looking at an animal in one particular environment isn't sufficient to guarantee that we have a true picture of its capabilities. "Nature plus nurture" suggests that the influence of either force can be teased out easily. But things aren't so simple. "Nature multiplied by nurture," or $P = G \times E$, is probably a better way to capture the complex phenotypic outcomes that are truly characteristic of animals.

WHY STUDY WORKING DOGS?

We've said that there are around a billion dogs in the world. Only a quarter of them live right in our homes or in very close association with us. Of these, many are meant only to be companions. Others, however, are animals that we primarily utilize for practical human purposes like agriculture. Our own research has been especially focused on these "working dogs."

Ray spent fifteen years breeding and training dogs that pull sleds, became a reasonably competent sled-dog racer, and wrote many papers on the anatomy, physiology, and mechanics of running dogs. Some four thousand dogs "went through the yard." After that, we both went on to study sheep dogs that assist farmers who keep livestock. We looked at two types: herding dogs, which conduct livestock from one place to another (Border collies from Scotland were our study animal), and livestock-guarding dogs that protect livestock from predators. We collected and studied Maremmas from Italy, Šarplaninac from Yugoslavia, and Anatolian shepherds from Turkey, along with lots of Chesapeake Bay retrievers raised and trained for "trialing" competitions by Ray's son Tim Coppinger. The final count on our livestock-guarding dogs was some fifteen hundred animals that we used for observational studies and experiments. Many of the guardian dogs ultimately went to cooperating farmers as working guardians across the United States and abroad.

They all turned out to be great animals to study if you want to know what makes dogs tick in general. Sled dogs were perfect models of animals selected for a single behavior—to run fast for very long distances pulling a sled and driver. And the working sheep dogs—guarding and herding—were an ethologist's dream subjects: two types of dog that were selected to work in the same environment with the same target animal but that differ strikingly in their patterns of behavior and the way they direct their behaviors toward livestock.

So let's look for a moment at how and why we got interested in these three kinds of working dogs. When he was younger and more limber, Ray decided he wanted to get into the arduous (but exciting) business of sled-dog racing. He acquired his first husky and showed it with pride to Charlie Belford, his local veterinarian and a world champion dog racer. Belford looked skeptical and asked, "How do you know she's a sled dog—has she ever pulled a sled?"

This seems like an odd question: huskies, after all, are the quintessential poster-child example of a sled dog. Now, sled dogs are an amazing animal—the fastest mammal in the world once you exceed a marathon distance. Nothing can touch them. For Belford, however, it clearly wasn't just a matter of being the "right breed." Something else was at stake. So these dogs provoked some great ethological questions: What is the relationship between genotype and phenotype in sled dogs? What are the adaptive characteristics that allow them to run in the Iditarod Trail race in Alaska for more than a thousand miles, under Arctic conditions, in eight days? What motivates them to do it? Can their particular biological machinery, their physiology and anatomy, explain the remarkable behavior of these animals?

We came to study livestock-guarding dogs in a very different way. Our interest in them began as an assignment—we had a rather practical goal in trying to understand their behavior. Since the early days of the American West, farmers and ranchers—and, later, government pest-control agents—relentlessly hunted, trapped, bombed, and poisoned wolves, coyotes, eagles, and pumas. Anything that preyed on domestic livestock was a target. By the early 1970s, lethal methods of predator control were beginning to be outlawed on federal lands throughout America. Wildlife biologists and conservationists were convinced that balanced ecosystems needed their top predators. Ecosystems weren't very well understood at the time, but when wolves were reintroduced into Yellowstone National Park after the mid-1990s, it became clear that species like the wolf do indeed play an astonishingly important role in the natural scheme of things. Before the wolves came back, elk populations had exploded in Yellowstone, and, over many years, vast numbers of grazing elk and bison had eaten much of the park's vegetation, especially along the rivers and streams where ungulates congregated to feed and drink.

The reintroduction of wolves changed everything again. As the browsers altered their grazing patterns to avoid the new pressure of wolf predation, grasslands and forests began to recover rapidly and dramatically along the rivers and streams. The regrowth of forest led to an explosion of bird diversity. Beavers returned, building dams with the resurgent trees and providing new habitat for a host of other aquatic animals that had long been gone from the park. The rivers themselves changed course (as writer and environmental activist George Monbiot put it in a compelling TED talk video about the impact of wolf reintroduction) when feeding pressure was reduced on the vegetation of their banks. Trying to understand the role that behavior plays in causing these big interdependent consequences of small changes in a complex biological system has had a profound influence on the thinking of ecologists, evolutionary biologists, and ethologists alike.

Even as predators like the wolf and the coyote were protected, the livelihoods of sheep and cattle farmers were still threatened. Some means of control that wouldn't be lethal to top predators was essential, and many such methods were investigated. One group of behavioral researchers in California, for example, tried training coyotes not to eat sheep by feeding them a nontoxic substance that tainted the taste of sheep meat—aversive conditioning. It worked pretty well but never caught on with the ranchers.

In our ethology and cognitive science program at Hampshire College in Amherst, Massachusetts, we explored a different approach. We had heard that pastoralists around the world used domestic dogs to deter predation on livestock. Almost everywhere—except in the United States—people had long used "special" guarding dogs whose only job was to protect farmed animals from predators: everything from lions and leopards to jackals and troops of baboons. In some Mediterranean countries, pastoralists used dogs to guard their animals from more familiar predators such as wolves, bears, and even other dogs, themselves a huge depredation problem in both this country and Western Europe. In some places, dogs were used to protect sheep from theft by people.

How did this kind of system work? What is it about these dogs that deters predators? What sort of behavior is required to enable a dog to peacefully coexist with sheep? Would they work in the United States to protect livestock in areas where predators have been restored? Studying the livestock-guarding dogs took us around the world. We found countries from Portugal to Italy to Turkey to Tibet that had one or more "landrace" populations of working dogs (local, naturally adapted varieties of dogs) that are sometimes pridefully called breeds (intentionally "designed" products of artificial selection). Written records suggest that people in these cultures had used landrace livestock-guarding dogs for several thousand years.

By the 1930s some of them were also being raised by dog fanciers in America and Europe—bred as pets and household guardians rather than working livestock-guarding dogs.

But we wanted to be able to study the original "natural" working stock, before breeders got hold of them, to find out how they actually did the job. Were they truly effective in preventing predation, or was this just another part of the robust mythology that dogs are man's best friend, always happy and willing to do our bidding? And if they were indeed successful, *how* did they work? We were especially intrigued because many domestic dogs will chase and kill sheep and other livestock, sometimes seemingly just for fun.

Our third study animal was the Border collie, a herding (or conducting) dog with a very different job: they are asked to guide the movement of large flocks of sheep by responding to commands given by a human shepherd. This is a dauntingly challenging thing to ask of any nonhuman animal, and no other type of dog is so widely used to intricately control the movement of livestock. Their prowess (in collaboration with a canny shepherd) is legendary, and if you've ever watched a highly competitive sheep-dog trial you'll appreciate the apparent skill and responsiveness to human command that it seems to require. What suits them to this job and style of work? Is it a product of painstaking training? Is it due to their native intelligence (they're sometimes regarded as among the "smartest" of dogs)? Or is their ability an intrinsic behavioral characteristic of the breed?

As we began our study of livestock-guarding dogs, we were frequently asked if these dogs could also be trained for the job of herding and work like Border collies. We didn't know the answer at the time—but it was a great question for ethologists and their students. How do these different working behaviors arise? Are there limits on

FIG. 5 THESE SIX-WEEK-OLD PUPPIES WERE ALL BORN ON THE SAME DAY; AT THIS AGE THEY ARE ABOUT THE SAME SIZE. LIKE MANY LIVESTOCK-GUARDING DOGS, MAREMMAS (*WHITE*) WILL END UP IN THE EIGHTY- TO HUNDRED-POUND RANGE BUT BORDER COLLIES (*BLACK*) WILL TOP OUT AT THIRTY-FIVE POUNDS. PHOTO BY LORNA COPPINGER.

what a particular type of dog can do? This was science at its best: along with observation of working behavior in the field, we could also conduct carefully controlled experiments to address these questions. For example, we could "cross-foster" puppies from each of the types that were specifically bred for one task or the other: each could be raised in the other's environment and trained to herd and guard in the same way.

It was a dream project. We (a group of Hampshire College students and their professors) traveled to the great sheep pastures of Mediterranean Europe and Asia with the goal of bringing back a breeding stock that could be used to produce a population large enough to be systematically studied on farms and ranches in the United States, as well as back in our lab. We collected livestock-guarding dog pups in Turkey, Yugoslavia, and Italy, carefully trying to buy pups in each of those countries that had been born on the same day (fig. 5) so we could control not only their developmental environment but also their rearing and training. On the way home, we stopped in Scotland and went to herding-dog trials, talked with handlers and farmers, and finally bought six dogs from working stock; four of them had been born on the same day as the livestock-guarding dog pups that we had just acquired.

Over the years, data from our lab and from cooperating farmers across the country have helped us learn a great deal about how "nature" and "nurture" play out in the behavior of these dogs. Further on, we'll discuss what we found in more detail, from many different perspectives, about all these dog varieties. For the moment we'd like to take a brief closer look at one small observation of a livestock-guarding dog behavior "in the field," and then explore how ethologists go about the work of describing and explaining it.

LIVESTOCK-GUARDING DOGS IN THE ABRUZZI

Our first assignment was to actually find guarding dogs in action and to describe their behavior in their natural setting. We had to develop a good definition of "guarding" so we could observe and measure it. We wanted to understand how a dog actually interacts with a flock of sheep; we wanted to know if their ability to peaceably coexist with livestock is a genetically programmed intrinsic characteristic of their breed or a result of training. And, important for the practical task at hand, we needed to know if guarding dogs were likely to kill a predator when it threatened sheep. After all, people tend to think of "guard dogs" (like snarling German shepherds on the perimeter of a prison camp) as animals big and ferocious enough to be capable of killing wolves or bears. If that was the story, those of us who were looking for a nonlethal method of predator control certainly wouldn't want to use these dogs—the whole point of the exercise was to find a method that discourages predation but leaves the predator intact.

The Abruzzi mountain range in east central Italy was a great place to investigate this interesting biological system "on the ground." (It was also a very desirable study area because the food in Italy is great—something every ethologist on a field assignment should consider when picking a study site.) In the Abruzzi, like in almost every traditional pastoral environment around the world, you'll find dogs living in and around the sheep. Countless generations of sheep have grazed on the Abruzzese hills and grass-rich pasturelands, sometimes with dogs and human shepherds, and sometimes in just the company of dogs. You rarely ever see a flock without dogs.

Like all animals—including dogs—the behavior of sheep is guided by the need to forage, avoid dangers like predators, and successfully produce offspring. That includes the predators themselves, of course: the wolves, foxes, lynx, and wild boars that still roam the Apennine Mountains also have to feed, reproduce, and avoid hazards. One way for these species to forage is to kill and eat sheep or their lambs. The story is, of course, different for the livestock-guarding dogs themselves; there is little for them to eat high up in the mountains—and like all dogs, they rely on human-provided food.

As ethologists we want to understand what makes this complex interaction of humans, sheep, dogs, and predators tick. So one summer morning early in our research, we met up with a floppy-hatted shepherd, some 350 sheep, and several large white Maremmano Abruzzese (Maremma) dogs. The sun can be infernally bright at noontime in these high elevations. Sunlight all by itself can be a killer of people and animals alike—only "mad dogs and Englishmen" go out in the noonday sun, and skin cancer and heat stroke are a threat for both dogs and shepherds. Hot dogs that can't find shade will pant and their pink tongues can get sunburned; indeed some of them develop cancers of the tongue. Big animals caught out in the sun can quickly die. Human shepherds are clever enough to protect themselves with appropriate clothing, as shown in figure 6.

On one particular exceptionally hot day the sheep slowed down and took a break at high noon. The dogs found a little shade under overhanging rocks; the shepherd crawled under his great field coat, which, along with his hat, protected him from the sun. He soon fell sound asleep as we sat and watched. After a little while, as the shepherd slept, his sheep got up and wandered off—and the dogs, in spite of the relentless hot sun, went off with them. After the animals had been completely out of sight for some time the shepherd woke up to discover that

FIG. 6   SHEPHERDS AROUND THE WORLD DRESS TO PROTECT THEMSELVES FROM THE
SUN. THE GREAT-COATED SHEPHERD ON THE TOP IS FROM THE PUSZTA GRASSLANDS OF
HUNGARY. BELOW IS ANOTHER SHEPHERD IN A GREATCOAT, FROM LESOTHO IN SOUTHERN
AFRICA. PHOTO BY LORNA COPPINGER (*TOP*) AND TIM COPPINGER (*BOTTOM*).

his flock and dogs had disappeared. He looked at us, a bit embarrassed, and we pointed out the direction in which the animals had gone. He waved a thank you and went off after them.

THINKING ABOUT MAREMMAS LIKE AN ETHOLOGIST

How do ethologists look at a little story like this? Well, science is really no more than a systematic way of asking and trying to answer questions. The challenge is to find the right kind of questions to ask and the right method for investigating them. The real trick is to formulate testable questions—hypotheses, or informed good guesses—that make predictions. Then we need to design field studies and experiments yielding data that allow us to check the validity of those predictions. Here is a sample of the many, many kinds of questions ethologists might ask about our one small observation of the behavior of Maremmas in the Abruzzi:

- Why did these dogs leave their human master behind (aren't they man's best friend?) and follow sheep instead?
- Would all Maremmas behave like the Maremmas we observed?
- Would any breed of dog act this way? How is this sort of working-dog behavior related to what our familiar household pets might do?
- Was their behavior a result of the sun and heat (or other features of the environment) in their high mountain habitat? Could they have been following sheep just to stay cool?
- Were the Abruzzi Maremmas intentionally trained to pay special attention to sheep and to follow them when they moved?
- What would happen if Maremmas were asked to protect goats or horses instead of sheep?
- Were these dogs just intelligent enough to figure out what needed to be done?
- Are livestock-guarding dogs born with a genetic preference for sheep over humans?
- Could the age of these particular dogs have made a difference?
- Was their behavior the result of something that happens as these

dogs develop and grow? If so, was there a particular time in life when this kind of relationship develops?

You shouldn't be dismayed to find that there could be so many different (and sometimes contradictory) ways of looking at and questioning one simple piece of dog behavior. When scientists investigate any kind of natural phenomenon it's almost always possible—and in fact desirable—to come up with multiple hypotheses. The heart of doing science is the creative activity of framing hypotheses and then testing them against carefully made and measured observations about the world. Arguing about and testing competing hypotheses is a great part of the fun and excitement of science.

And there's another, a more fundamental reason why different scientists may come up with different explanations for why things are the way they are. When we ask why a particular behavior occurs, we're really not asking a simple question. Niko Tinbergen, one of the founders of modern ethology, argued that it's necessary for biologists to think about the "why" of behavior in at least four different ways. These have become known, famously among ethology students, as Tinbergen's four questions.

Here is how Tinbergen himself posed these questions in a 1968 article in the journal *Science*:

1 In what way does this phenomenon (behavior) influence the survival, the success of the animal?
2 What makes behavior happen at any given moment? How does its "machinery" work?
3 How does the behavior machinery develop as the individual grows up?
4 How have the behavior systems of each species evolved until they became what they are now?

The first question is about the function of a behavior: how does it meet the fundamental needs of an animal to find food, avoid hazards, and reproduce? When the Abruzzi Maremmas followed sheep but not the shepherd, did that behavior enhance their ability to feed or leave more or better offspring than guarding dogs that don't stay with the flock? Could the simple answer to our puzzle be that shepherds feed and support those dogs that follow sheep and cull those that don't?

Second, Tinbergen asked us to inquire about mechanism (or causation). Along with the biomechanics of movement itself, what other physiological, neural, and motivational processes might or might not be at play when an animal actually moves in space and time? Were our Maremmas influenced to move by the direct physical effects of high temperature? Is there a neural circuit "wired" in their brains that is triggered by the movement of sheep? Are they internally rewarded because it simply feels good to follow sheep? When they do it, are their brains providing a pleasurable slug of endorphins like a runner's high?

Third, we have to look at the development (or ontogeny) of a behavior: When does the behavior appear in the life history of an animal? Is it stereotypical and unalterable (intrinsic), or does it change over time? Do livestock-guarding dogs that grow up with sheep at some critical time in their development simply "mistake" sheep for other dogs that they would prefer to stay with? Would the same individuals, if raised with goats, not follow sheep at all?

Fourth and finally, and of critical importance to ethology, we need to understand the evolutionary history (or phylogeny) of a species. Why did the course of biological evolution give rise to a particular way of acting? What was the adaptive value of a behavior pattern on which natural selection operated? How does behavior play out in related species that share an evolutionary past? Earlier, we said that dogs aren't wolves; but they do have a common ancestry and a comparison of behavioral differences between them (we'll do this frequently as the book proceeds) is likely to be illuminating.

We will come back to all of these questions about Maremmas (and similar questions about sled dogs and Border collies) later on in the book. For the moment, let's look at a different but familiar kind of animal and a very "simple" question — the kind a child might ask — whose answer, as Tinbergen suggests, is not so simple.

"Why do birds fly?" From the viewpoint of mechanism, of immediate causation, the answer is that flight results from the activation of a set of skeletal and muscle movements in a pattern that generates aerodynamic lift of wing structures. To what functional end, though? It's a good bet that birds do it because flight improves their chances of survival and reproduction; this form of locomotion allows them to escape from predators relatively easily and to carry out long-distance searches for food and shelter. The evolutionary history of birds provides an answer from yet another viewpoint: they now have feathered wings that allow flight because their ancestors evolved that kind of structure. Wings may in part have had a different original function — it is possible that the dinosaur ancestors of birds evolved feathers simply for insulation or to attract mates: in the end, the forces of evolution provided birds with genes that built an aerodynamic shape that is capable of flight. Finally, we can say that birds fly because they develop in a particular way. Newborn fledglings in the nest (let alone eggs) can't fly. Their body shapes, feathers, and nervous systems are adapted to nest living, and they need time to metamorphose into adult birds. Indeed, some birds need to practice before they can do it well. So why do birds fly? There's no single answer. It's because flight serves a functional adaptive purpose (or purposes) in their lives; it's because they have shapes and physical mechanisms that enable it to happen; it's because their evolutionary history provided the means; it's because they change and develop over a lifetime, and their bodies (and brains) change shape in ways that make flight possible. In order to come to a full understanding of the behavior of a dog, or any other mammal, we ultimately need a multidimensional and integrated synthesis of insights at all of these levels.

3 THE SHAPE OF A DOG IS WHAT MAKES IT TICK

THE SHAPE OF SLED DOGS

People often say that dogs have some of the most varied shapes of any species—just picture a Chihuahua, a Great Dane and a bulldog. While they may not be the world champions of variation—pigeons and chickens exhibit a spectacular range of colors and designs as well—dogs certainly do come in a remarkable panoply of sizes and shapes. Many of these are thought to be adaptive in breeds that are used (and were bred) by humans to perform specific tasks. We begin our discussion by looking at the sled dog as an example of the fundamental importance of size and shape.

Working draft animals with many different shapes have been used by humans for traction—to overcome friction and pull things more efficiently than we can ourselves. Horses and oxen pull plows; donkeys pull carts. Occasionally you'll see a small wagon pulled by a pig—or even a pair of pigeons trained to pull a toy vehicle. Of course, the size and structure of the device that is pulled needs to be appropriate to the shape of the draft animal. Oxen may be able to pull a few tons of concrete in a contest at the county fair—and might even outcompete mechanical tractors—but you can't expect the family dog (let alone a pigeon!) to do it.

Dogs, however, are great at pulling sleds over snow. People in northern climes have used them for transportation and haulage for centuries. Today you're most likely to see this behavior in the sporting context of sled-dog racing. Every year since 1973, for instance, one of the most strenuous athletic events in the world takes place in Alaska—the 1,100-mile Iditarod sled-dog race from Anchorage to Nome. Dozens of human mushers and hundreds of dogs take part in the competition, which has a record time of eight days and fourteen hours. The teams are running six marathons a day for over a week. This is, to say the least, a demanding behavior for any kind of animal: what makes it possible for (some) dogs to do it? The answer is that they have to have just the right shape (fig. 7).

People are often surprised when they look at a sled-dog team and find it composed of dogs with a single shape. On a good team, the pairs of dogs are almost mirror images of each other, so that their reach and gait are matched. But they don't have the shape you might expect. The team isn't a dozen or so big and robust animals of a single breed like the Alaskan malamute or the Siberian husky, canonical examples of hardy northern dogs. Often, today's sled dogs are hybrids or mixed breeds that are purposefully bred to the kind of genetic standard you see in modern hybrid corn, where agriculturalists strive to produce plants whose ears are exactly the same size and at the same height from the ground in order to facilitate mechanical pickers. Sled dogs as a group are sometimes called Alaskan huskies, as if they were a single kind of dog, but in truth they can look, in many respects, like a pretty motley crew. What really matters—and this is the way in which they resemble one another—is that sled dogs are of the right size.

Racing sled dogs all weigh in at about fifty pounds. Chihuahuas, it may go without saying, make poor sled dogs. They're far too small and simply can't pull their weight. The canonical Alaskan malamutes are in fact also the wrong shape for a modern racing team, because they are too big. Breeders sometimes boast that they can produce bigger and bigger malamutes; some are over a hundred pounds. When dogs get up toward sixty pounds, however, old professional sled-dog drivers begin to raise their eyebrows. At the starting line of a sled-dog race, the chief judge will often look a team over and take a dog off if he doesn't think it can perform—and the chief reason to eliminate a dog from competition is its weight. An overweight animal has the wrong shape. You'll soon see why. Winning teams of racing sled dogs are composed of individuals with the best uniform size and shape for the task at hand. If you

FIG. 7   THE SHAPE OF A WINNING SLED-DOG TEAM. RACING AFICIONADOS MAY WANT TO NOTE THE LEAD DOG IS PUMPING WITH HIS FRONT FEET BECAUSE HIS SHOULDERS ARE SORE, AND THE DOG ON THE BACK LEFT HAS PULLED HER HEAD AND TAIL UP BECAUSE THE SPEED IS FASTER THAN SHE WANTS TO RUN. THE DOG ON THE RIGHT-HAND "WHEEL POSITION" BEHIND THE LEADER HAS ABOUT AS PERFECT A GAIT AS YOU CAN GET. PHOTO BY LORNA COPPINGER.

FIG. 8 MORE WINNING SHAPES. AT TOP SPEED, THE DOGS RUN SEEMINGLY EFFORTLESSLY AND ARE ALMOST PERFECTLY MATCHED IN PHYSICAL SIZE, SHAPE, AND MOVEMENT PATTERN. THE LAST DOG ON THE LEFT SIDE (*ARROWS*) IS TIRED AND THE DRIVER IS "WHISPERING" GENTLY TO HIM, "GOOD BOY." THE RACE DEPENDS ON NOT INCREASING THE SPEED BUT THAT DOG MAKING IT FOR THE LAST MILE. PHOTO BY EMILY GROVES YAZWINSKI.

change the size and shape of a sled dog, the animal will be unable to move properly for the task at hand—and it will be very uncomfortable trying (fig. 8).

MEASURING SHAPES

To understand why this is so, we need to first look more closely at what we mean by shape and how we can measure it. "Shape" really is a cover term for the myriad physical characteristics of a biological machine— its size, color, skeleton, brain and nervous system, all of its other organ systems, and ultimately its cells and cellular products, such as hormones. Everything and anything about the shape of an animal can play a role in its behavior—which you will recall we have defined as "how the animal's shape moves in time and space." And all the shape properties of a sled dog come into play when we try to figure out what makes it tick—what makes it able to behave like a "good" racing animal. Let's begin by looking at a simple and easily observable property of shape: the overall external form of the machine.

An organism, like any object at all, is three-dimensional: it is so high, so long, and so wide, and we can measure its volume, weight, and surface area. Real animals (and machines), of course, usually aren't just simple geometric objects and they often have remarkably complicated shapes. For the moment, however, let's imagine a breed of "dog" that is shaped exactly like a spherical ball and a second breed that is shaped exactly like a perfect cube, such as in figure 9.

We'll assume that these two "breeds" are identical in volume: they occupy the same amount of space and have the same number of cells. They are, of course, dissimilar in shape in other ways. For their volumes to be identical, they will have to differ from one another a bit in width and diameter or in end-to-end size. We'll need to review some high school geometry here. You may recall that the formula for the volume of a rectangular solid is $V = lwh$ (i.e., its length × its width × its height). If we look at a "cube dog" with a width and height and length of 2 units (inches, centimeters—pick your favorite unit) on all sides, then its volume will be $2 \times 2 \times 2 = 8$. How big would a "ball dog" have to be so that it also has the same volume of 8? The formula for the volume of a ball

$D = 2.48$ $W = 2.00$

FIG. 9 THE "CUBE DOG" AND THE "BALL DOG" ARE THE SAME SIZE (THE SAME VOLUME AND THE SAME NUMBER OF CELLS). BUT THEIR SHAPES DIFFER AND THEY WILL HAVE TO "BEHAVE" VERY DIFFERENTLY. DRAWING BY PETER PINARDI.

is $V = 4/3 \, \pi \, r^3$ (4/3 × the value of π × the radius, cubed.) So when we do the calculation, $4/3 \times 3.14 \times r^3 = 8$. After we do the math to solve for r^3, we find this value is 1.91. The radius of the ball, r, would be 1.24. The diameter of our ball would be $2r$ (two times the radius), 2 × 1.24, or 2.48 units. With all our calculating done, we have a cube dog with a side-to-side width of 2; it has the same volume as a ball animal with a diameter of 2.48, or 20 percent greater than the cube animal's width.

If one of our geometric dogs needs to pass through a doorway with a width of 2 and a bit, it had better be cube shaped. A ball-shaped dog wouldn't make it through the door, even though, in a sense, the two breeds are the same size. Put another way, these simple shape differences constrain a ball dog to act differently in the world than the cube dog will. There will be some movements that the cube dog can carry out that the ball dog can't. No amount of training is going to get the ball dog through the door.

From a slightly different point of view, think of these shapes as two different species living in the same habitat; they share an environmental niche. Whether an animal is shaped like a ball or a cube will matter a great deal in its ability to negotiate a common environment. Different shapes will have different energetic consequences as well and will directly affect how the species are able to move—to behave. Suppose the two species share a hillside habitat. If you put a ball at the top of a hill and impart energy (push it), it will roll down to the bottom all by itself. It takes very little energy to start it, and the trip down is practically free because gravity helps it along. If you put a cube species in the same place and push it, it's not likely to move much at all, let alone roll. If the hill were steep and smooth and slippery enough, the cube might slide down — but it wouldn't roll. It might make it all the way to the bottom but it would be slower than the ball because there would be more surface area touching the ground, creating more friction and slowing the descent. The cube would almost certainly never get to the bottom of the hill as fast as the ball. The cube would have to put a lot more energy in getting to the bottom, which means, at the very least, it would have to eat more than the ball to get the necessary energy to accomplish the same task even though they have the same volume and the same number of cells.

Dramatically different behavioral outcomes result from these differences in shape. A ball will rarely slide without rotating and a cube will rarely tumble end over end. Whether you would rather be a ball or a cube depends on the particular features of the world around you and the behavioral requirements they impose. If all the available food for the two species was at the top of the hill, a ball animal would be at a disadvantage because it would risk constantly rolling away from the food or spending energy trying not to roll away from it. The cube species would be able to just sit there like a giant sloth, expending hardly any energy feeding. At the same time, if predators showed up at the top of the hill, it would be great advantage to be a ball, able to roll off more quickly and efficiently than a cube ever could. Like life, shape is always a compromise.

Now let's look at our hypothetical ball and cube dogs from another mathematical perspective. What would their surface areas be? The surface area of any one side of a cube is its height (h) × its width (w). We then multiply by 6 (since a cube has six sides). So for our cube dog, with each square side having a height and width of 2, the surface area calculation is $h \times w \times 6$, or $2 \times 2 \times 6 = 24$ (square inches or whatever your favorite unit is).

As for a ball dog, the surface area equals $4 \pi r^2$. For our particular ball dog, that's $4 \times 3.14 \times (1.24 \times 1.24) = 19.2$. Our cube dog with the same volume (8) as our ball dog has 25 percent more surface area than the ball animal does. Now you might think that if two animals have the same volume—hence, the same number of cells—they would require the same amount of food (ultimately, sugar) to run their cells, regardless of the surface area. You have to remember, however, that metabolism, which runs the cellular machinery that burns sugar to produce energy, also generates heat. If you think of the surface area as a radiator, you'll begin to see the point. When a dog is working it generates extra heat because it is has to burn extra sugar to provide extra energy for the work. The dog has to get rid of that increased heat—otherwise its temperature will rise. A dog has a normal temperature of 101.5°F; its mechanism is built to work best at that temperature. When an animal's temperature gets too high, its cells will die. So heat has to get out of the organism, and in dogs much of the extra heat is radiated off the animal's surface. How well a dog can do that depends largely on how much surface area it has (and on the temperature of the external environment).

So if we have our two "geometric" dogs pulling a sled—working hard and creating a lot of body heat—then (all other things being equal) the cube would make the better sled dog. Running a race generates an enormous amount of heat, and the cube dog will be better able to get rid of it simply because for the same volume it has more surface area than the ball does.

But as we've lamented, life and shape are always a compromise. If our two dog animals weren't working as sled dogs generating too much heat but were simply living in the cold Arctic, they would have a different problem—how to conserve heat in order to maintain a constant normal temperature. Which shape would be better suited? In a frigid climate, it would be the ball dog with the smaller surface-to-volume ratio. Yes, our cube animal would make a better sled dog with its large radiating surface, but at rest it had better have a good well-insulated doghouse and a blanket or it will freeze to death by radiating off all that heat when it is not working.

Dog people love to form breed clubs and to compete at dog shows to see who has the best animal. A veteran of these competitions once said that "all other things being equal, the judge will give the ribbon

TABLE 1. SHAPE AND SIZE

	Little Ball	Big Ball	Little Cube	Big Cube
Surface area	19.3	77.3	24.0	96.0
Volume	8.0	63.7	8.0	64.0
Surface/volume ratio	2.4	1.2	3.0	1.5

to the larger dog." Over the years show dogs, which don't work, get steadily bigger (all except Chihuahuas, which are specially prized as miniatures). Does that mean that bigger animals are always better? Are bigger sled dogs better? We need to look at this question quite closely.

Let's imagine that there were breed clubs and shows for cube and ball dogs. Suppose that judges follow the dog-show fashion and give preference to bigger cube dogs with, say, twice the side length of the cube animal in the figure, and to bigger ball dogs with twice the diameter. These bigger animals may get the blue ribbon, and as aficionados go on to breed champion to champion, our average two geometric dogs will at some point double in size. Let's say that a typical champion ball dog now has a diameter of 4.96, and a prize cube dog's edge will now be 4. What happens to their surface areas and their volumes? Remember your high school geometry formulas and get out your calculator:

Volume of a larger "champion" ball = $4/3 \times 3.14 \times 15.25 = 63.7$

Surface area of the ball = $4 \times 3.14 \times (2.48 \times 2.48) = 77.3$

Volume of a larger "champion" cube = $4 \times 4 \times 4 = 64$

Surface area of the cube = $4^2 \times 6 = 96$

Table 1 sums up the magnitude of size and shape differences between smaller and larger ball and cube dogs. Doubling the size of both breeds has cut their surface-to-volume ratio in half. Dog show judges might prefer larger animals, and breeders for show might favor them, but the bigger animals are not going to radiate heat as well as their smaller relatives. What our imaginary judges and breeders haven't considered is that to change size is to change the overall shape—and when it comes to behavior in the real world, shape matters a great deal. Indeed, in living organisms, as the great British biologist J. B. S. Haldane once observed, "a large change in size inevitably carries with it a change of form."

If an animal's challenge is to avoid losing heat to the atmosphere in the cold north, for instance, then of the four shapes, the big ball has the advantage: it has the smallest surface-to-volume ratio and (all things being equal) it will radiate the least heat. The big cube would be the next most preferable shape, then the little ball, and last, the little cube. In general, pound for pound, balls are better at adapting to cold than cubes. Is it any wonder then that many animals that live (or have lived) in the Arctic, such as mastodons, polar bears, whales, and walruses, are large, rotund, and stocky—in effect relatively giant balls? Is it any wonder that on cold nights, dogs and many other mammals change their shape by rolling up into balls, which reduces their radiative surface area?

Aside from just radiating heat, another benefit is that a big ball won't need to eat as much per cell as a little one. For its size, the big ball-shaped animal doesn't give up as much energy in heat: overall it is more energy efficient than the small one. So the effect of these shape differences on the animal's economy can be profound. That is what natural selection is all about. Small differences in shape are selected for or against. Those changes in shape that are more energy efficient are selected for: they get to leave their offspring to the next generation more often.

Sled-dog racers may not ever think explicitly about these theoretical considerations. They select for dogs that can win, to be sure, but that really means they have selected for dogs with the best shapes for the energetic demands of racing—animals that don't overheat and will have adequate energy reserves. Why is any typical fifty-pound dog a better racing specimen than a hundred-pound malamute? If malamutes and racing sled dogs are anything like ball dogs and cube dogs, then the hundred-pound malamute has twice the volume and 60 percent of the surface-to-volume ratio of a good racing dog. That is, the malamute has twice the number of cells generating enormous heat at racing speed, but less than half of the surface-to-volume ratio capable of shedding that heat. Moreover the malamute has to carry around fifty extra pounds as it races: this is an additional energy cost in itself, which will require burning even more calories and creating yet more heat.

So why not use two-pound Chihuahuas with one twenty-fifth the weight of a typical sled dog and the greatest surface-to-volume ratio among all the dogs? Well, there is clearly more to the behavioral impact of shape and size than energetics alone. A team of racing Chihuahuas won't have the overheating problem of a malamute. However, their small size can't provide the muscle mass and action that is required to pull a heavy sled. When they lean into the harness they are so light that the sled won't move. Furthermore, when they run, each Chihuahua step will cover just a tiny distance compared to the bigger dogs with greater reach. In this regard, the malamute's much greater mass certainly makes it a better candidate for pulling a sled, and a big malamute will have the reach to cover a lot of ground with each stride. If muscle mass and reach were the only shape considerations, malamutes would make for a great sled-racing dog. Unhappily for a malamute with aspirations to win Alaska's Iditarod Trail Race, it has the wrong shape for proper cooling. The malamute will have to run more slowly. The shape difference is the critical factor limiting its behavior.

Weight has other consequences as well. When 115-pound Frank Shorter won the Munich Olympic marathon in 1972, he reportedly sat up the night before and sandpapered his sneakers to reduce his weight-to-shoe ratio by four ounces. Four ounces may not sound like much, but when you multiply the energy cost of picking up four ounces times the number of steps in a marathon, the total weight lifted is enormous. For the same reason, a sled-dog driver chooses dogs that are able to skim along the surface of the snow rather than pop up and down. The "popper" is picking its body weight up over and over again for the whole race. This costs much more energy and produces much more heat than if the dog kept its body at a constant height throughout the race. So what makes one dog pop and another not? The shape of the pelvic girdle and the angle at which it is attached to the backbone—the position of its hind legs, the way the head and neck sit on the backbone, the length of the backbone, and the flexibility of the backbone to stretch with the front and back legs—all of these mechanical characteristics of a dog contribute to an overall shape that supports a particular kind of locomotion. The shape and resulting movement pattern of a "popper" is inefficient when it comes to long-distance racing.

At just about any race, some onlooker will ask if a driver has cross-bred his dogs with wolves. Aren't wolves simply stronger, faster, and overall better than their domesticated relatives? Wouldn't wolf genes improve performance? Certainly, adherents of the romantic mythology of the "wild" would like to think so. Now wolves average one hundred pounds in weight, and they live in the arctic and the subarctic. We already know that it is not to an animal's selective advantage to have a lot of radiating surface area if it lives in cold regions. So like big malamutes (originating in a similar climate), the hundred-pound wolf has a shape that is good for conserving body heat—but a wolf will "run too hot" to be good at running marathon distances in record time.

Let's look at this another way. Imagine a hundred-pound wolf that had to run three-minute miles for twenty-five miles, day after day. When dogs race on the Iditarod Trail they can burn 10,500 calories a day. Good deer meat (a staple of some wild wolves' diet) provides about seven hundred calories per pound. Assuming that the metabolic requirements of dogs and wolves are similar, if not identical, a hundred-pound wolf would have to eat thirty pounds of deer a day in order to race. In normal circumstances, large wolves probably eat and digest about five pounds per day on average. Even if the imaginary wolf could actually procure and eat thirty pounds of meat every day, it would probably suffer from acute indigestion! And with that much food in its stomach, a hundred-pound wolf would now weigh 130 pounds—which means it would have to spend 30 percent more energy carrying the food around while it digests it. Thus the feeding habits, activity levels, and behavioral possibilities of a wolf (or indeed of any animal) are limited by shape characteristics like size.

The same is true for dogs, of course—though they have the great advantage of human care. On long races like the Iditarod, dogs are in fact fed many times a day with a diet high in energetically efficient fats and oils. Their feed is ground up to provide lots of digestive surface, and it is heated to body temperature so the animal doesn't have to spend extra calories warming its food and water to 101.5°F. To put this all another way, an animal's behavioral potential is limited by the "economics" of its shape (as behavioral ecologists might put it). For any behavior, the profit derived from movement (the benefit) has to be greater than the cost of moving. Whatever benefit there might be for a wolf being able to run marathon distances is offset by an enormous caloric expense (thirty pounds of deer consumed every day). Otherwise natural selection would have made wolves 50 percent smaller.

The nonnegotiable behavioral bottom line for wild animals is that they have to find their own food (an energetically expensive activity in its own right) and catch and kill it—which also costs calories and can be dangerous: they can get hurt and may incur expensive "repair costs." If we add up the calories in a moose and divide that benefit by the cost in calories of finding and killing it (as well as repairing a body that may have been kicked by the moose) the number of calories a wolf has to retrieve from a moose hunt must be greater than calories it expends. In addition, wolves will need some extra energy for making and feeding puppies, a notoriously expensive output of calories. While they are finding food and having puppies, they also have to nervously keep their eyes open and be able to have enough calories to run away from bears that are trying to steal their food or eat their puppies. Accomplishing all three of these basic behavioral requirements is like a winning a race; the prize is successfully passing on your genes. Indeed for animals in general, life is like a sled-dog race—and most of the teams lose. For wolves, the energetic return from simultaneously managing all of its essential foraging, reproductive, and hazard-avoidance behaviors is often exceeded by the actual costs of moving—and they starve to death. It is true: most wolves starve to death. The sober fact is that the vast majority of animals that are born are not going to find enough food to sustain themselves energetically, will be eaten themselves, or will meet some other unfortunate end and will fail to breed.

From an adaptive point of view, then, working dogs have a very simple (if not idyllic) life. All they have to do is win an actual race, or a sheep trial, or stay out there in the field guarding sheep and wait for humans to provide them with enough food to perform those tasks. The best working dogs will be selected to reproduce, and humans will care for their offspring (which, as we've suggested, is a very expensive behavioral proposition for any animal). As a result, well-bred working dogs are able to afford to perform particular tasks efficiently even when the energetic costs are relatively high.

Because their reproduction is under human control, however, dogs that can't make the winning team, or do a job properly, often don't get to breed. Darwin called this artificial selection—differential reproduction guided by the decisions and interests of humans. In fact, Darwin first developed his revolutionary ideas about evolution by natural selection in the context of thinking about how humans have bred and shaped domesticated animals like the dog. His great insight was to realize that the two methods of selection, artificial and natural, were effectively the same kind of biological process. In his letter to Asa Gray, Darwin says:

> It is wonderful what the principle of Selection by Man, that is the picking out of individuals with any desired quality, and breeding from them, and again picking out, can do. Even Breeders have been astonished at their own results. They can act on differences inappreciable to an uneducated eye.... Selection acts only by the accumulation of slight or greater variations, caused by external conditions, or by the mere fact that in generation the child is not absolutely similar to its parent. Man by this power of accumulating variations adapts living beings to his wants,—he may be said to make the wool of one sheep good for carpets and another for cloth, &c.— ... Now suppose there were a being, who did not judge by mere external appearance, but who could study the whole internal organization—who was never capricious,—and should go on selecting for one end during millions of generations, who will say what he might not effect!

So a racing sled dog is not just any fifty-pound dog. It is an integrated machine whose "external appearance ... and whole internal organization," as Darwin put it, is admirably suited for the performance of pulling a sled at racing speeds. Simply by breeding the fastest animal to the fastest animal, the animals that can perform best in the special environment of sled-dog racing—it is behavior, after all, that drives evolution—have evolved a unique shape that is in fact not found anywhere else in the dog world, or even in the wider natural world. It has been a successful shape. When sled-dog racing began at the start of the twentieth century, winning teams were running five-minute miles. By the 1960s, you had to run a four-minute mile to win. Now the speed has increased to under three-minute miles for a marathon distance. That is evolution at work—whether it is driven by natural or artificial selection.

VARIATION IN SHAPE

When we talk about a species we usually think of it as having an ideal or prototypical shape. We point to a picture-book image and tell our children, "This is a lion." However, the notion that every member of a species is exactly the same is, of course, just a convenient fiction, even if it's a handy way for us to talk about the world. In fact, every animal in any population varies a little in its shape (and evolution by natural selection depends crucially on that variation). Two people, or two poodles, may look superficially similar. In truth, however, no two individuals are identically shaped, and as a consequence no two individuals will behave exactly alike. We'll look at this assertion in much more detail when we discuss how animals develop through their lifetimes.

It's also critical to remember that the shape of an animal is not simply its external form, its visible body. Animals also have extraordinarily complex internal shapes—the inner workings of the machine—and individuals as well as breeds (and species) can vary significantly in the details of those component structures, too. For example, one of our former students and now colleague, Cynthia Arons (along with neurophysiologist W. J. Shoemaker), compared the brains of sled dogs, guarding dogs, and herding dogs (see fig. 10). She discovered that in four brain regions these three types of working dogs differ significantly in the amount of dopamine found in the neural tissue. Dopamine is a neurotransmitter that is known to mediate general arousal and motor activity. Not surprisingly, frenetic and hyperactive Border collies, which are expected to chase after and change the direction of large groups of sheep, show as much as four times the dopamine level of stolid, slow-moving Maremmas, whose job is to simply to stay with a flock.

This local physical difference in brain tissue—a shape difference—has profound consequences for behavior and sets limits on potential behavioral plasticity, that is, learning. Just as you can't "teach" a cube to roll because the geometry of its shape restricts its movement—just

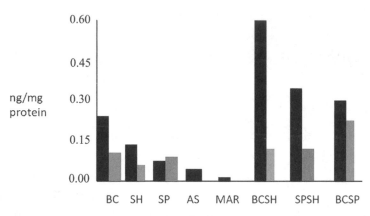

DIFFERENTIAL EXPRESSION OF DOPAMINE BY NEURONS IN
TWO BRAIN REGIONS (black = SEPTAL NUCLEI, gray = AMYGDALA) IN EIGHT DOG BREEDS

FIG. 10   THE SHAPE OF THE BRAIN IS VERY DIFFERENT IN WORKING AND HUNTING BREEDS,
UNDERLYING THEIR SIGNIFICANT BEHAVIORAL DIFFERENCES. DR. CYNTHIA ARONS
LOOKED AT SEVERAL BRAIN REGIONS TO MEASURE HOW DIFFERENT NEURAL STRUCTURES
IN THESE AREAS EXPRESS NEUROTRANSMITTERS SUCH AS DOPAMINE IN BORDER COLLIES
(*BC*), ŠARPLANINAC (*SP*), SIBERIAN HUSKY (*SH*), MAREMMA (*MAR*), ANATOLIAN SHEPHERDS
(*AS*), AND CROSSBREEDS BETWEEN THEM. NOTE THAT IN RELATIVELY PLACID ANATOLIAN
SHEPHERDS AND MAREMMAS THERE WAS *NO* MEASURABLE DOPAMINE EXPRESSED IN THE
AMYGDALA, WHICH IN MANY SPECIES IS INVOLVED IN AROUSAL AND AGGRESSION; NOTE
AS WELL THAT CROSSBREEDS ARE NOT THE AVERAGE OF THE PARENTAL BREEDS. GRAPH
ADAPTED FROM ARONS AND SHOEMAKER (1992).

as you can't train a dachshund to run like a greyhound—you can't teach
a guarding dog to herd sheep—because, among other things, its brain
shape doesn't support the right kind of motor activity.

Furthermore, no two individuals live and move in exactly the same
environment, and even small shape differences can play out differently
when they interact with variations in the physical world. Balls can, in
general, roll easily down hills because of their shape. But "your mile-
age may vary" when it comes to exactly how fast a particular individual
ball may go and precisely where it ends up. A tennis ball and a base-
ball of the same diameter have different weights and internal structure and they are built out of different materials. These two variations
on the general ball shape are likely to respond differently to a host of
environmental factors: the soil type on an incline on which they might
be rolled; the amount and height of grass cover; wind speed and rain;
or the random distribution of rocks. And, just as no two animals are
identical, no actual ball is perfectly round (our mathematical analogy

notwithstanding), and no two individual balls are imperfectly round in exactly the same way. When it comes to how they move and behave, small environmental differences can magnify that variability in subtle but significant ways.

All that said, however, the more physically alike are any two individuals, the more they will tend to behave alike. That's why we are able to talk about the behavioral characteristics of a particular breed whose members share a general shape. It is why we can refer to "species-typical behavior" or "breed-typical behavior" at all. It is why we can sum up this part of our story by saying that a dog behaves like a dog because it is shaped like a dog, and Border collies or Maremmas don't behave like other breeds because they are not shaped like other breeds.

4 THE SHAPE
OF BEHAVIOR

MOTOR PATTERNS

When animals are said to behave, we mean that they take on a particular body shape that occupies space and time. A show dog in the ring is taught to stand up with its feet symmetrically placed at its four corners, with its tail and head held rigidly at specific angles. That is one shape of a dog. In the abstract, an animal has a species-typical shape. It also has an actual individual shape that is measurable in terms of properties like weight. But most "behavioral shapes" in animals are dynamic: like the position of hour, minute and second hands on a ticking clock, the animal's physical configuration in space changes over time. At any given moment its intricate shape is capable of "behaving" in countless different ways. Some of those moving shapes were products of natural selection and are thought to be adaptive: the best shapes for feeding, reproducing, and avoiding trouble. Ethologists call them motor patterns.

Consider your own hand: it occupies a different space during every second of your lifetime. It might never make exactly the same movement twice in the same space ever again. But if you look for the highest frequency of hand motion in any population of people, you will see them using the hand to lift food to the mouth. That pattern may vary a little bit from person to person at any given meal, but the general form of the movement is a characteristic aspect of human behavior. Wherever you go in the world, whatever culture you visit, whether fingers, chopsticks, or knives and forks are used, everybody feeds themselves with their hands. The moving shape of hand to mouth is an adaptive product of evolution that supports foraging behavior—it is a species-general motor pattern.

Like any motor pattern, the behavior is, in large measure, a function of the physical shape of the organism's machinery. The particular structural configuration of the hand—its finger bones, joints, muscles, nerves, and skin—is essential to the task. So is the mechanical linkage between hand and arm and the wiring in the motor cortex of the brain that drives its movement: when you bend your elbow from a resting position at your side, the hand lines up perfectly (adaptively) with the mouth. In effect we may well say that the intrinsic shape of the human machinery defines the rules that govern our feeding behavior. In the same way, the shape of a dog plays a critical part in defining the rules for dog foraging and feeding. We'll further explore this notion of rule-governed behavior, and what we mean by "rules," in a short while.

Can development and learning affect hand shape and behavior? Yes, but only to a limited extent. The shape of the human hand is indeed central to the self-feeding motor pattern, but of course it also plays a role in a multitude of remarkable human activities, from using tools to playing the saxophone. One pair of hands can be trained to play the piano with virtuosity; another makes a fist more frequently than others; another is callused from hard work. That said, no human hand can grasp an object much larger than the extended span of its fingers. The shape and size of a hand, and its range of movement in space and time, set pretty severe limits on the total set of movements that dexterous humans might otherwise want to—or be adapted to—display. Remember that ethologists won the Nobel Prize for saying that "behavior is a taxonomic characteristic of a species." Just as a hand with five fingers is a product of evolution that characterizes our species, so, too, is the behavior of our five digits—and so is the pattern of movement of those fingers as they deliver food to the mouth.

This perspective provides ethologists with a straightforward way to describe behavior—do it by measuring how and when a hand, or a foot, or a whole dog moves in time and space. But which movements should we measure? What are the "differences that make a difference"? Should we take account of twitch of a little finger? Every eyeblink? Any small tilt of the head? The ethological answer is to focus on those movements that we think are adaptive products of evolution—patterns of behavior that arose because they increase an animal's fitness, its ability to feed, avoid hazards, and reproduce. They are taxonomic traits—present in all members of a particular species and often generally present in related species that share an evolutionary history. They are genetic, arising from the action of an inherited genome that guides the construction of the machine. And their overall character is stereotyped: they are essentially the same in every individual. Lorenz and Tinbergen called these motor patterns or fixed action patterns.

Like "innateness" and "instinct," the term "motor pattern" isn't too fashionable these days either. For some contemporary behavioral scientists the phrase may suggest a simplistic picture that overlooks the importance of learning, development, and the influence of the environment. In the early days of ethology, motor patterns were referred to as "fixed action patterns," suggesting a higher degree of stereotypical uniformity in movement and behavior than most animals actually exhibit. Later ethologists came to realize that completely "fixed" movement patterns are relatively rare. George Barlow, who studied behavior in cichlid fishes, subsequently coined the term "modal action pattern" to capture the idea that motor patterns are best thought of as reflections of average behavior. Moreover, in the early days the focus was indeed on simple single actions that occur essentially instantaneously in time. It's now clear—we'll see good examples of this shortly—that some motor patterns are actually complex chains of events that may play out over a considerable period of time. Our view, however, is that, along with many ideas that are due to the old ethologists, the basic concept of the motor pattern remains fundamental in understanding some very important dimensions of the behavior of dogs and other animals.

To see why this is so, let's take another look at the working dogs—animals used for guarding and herding livestock, sporting (like our sled dogs), hunting, military dogs, assistance dogs for the blind. In contrast with companion or pet dogs—whose behavior is often deemed "good enough" if the animal is housebroken, doesn't bite, and comes when called—working animals have to satisfy their owners that they are able to carry out very specific tasks in an effective way. People who breed, train, and manage working dogs often have lots of them, and they recognize that not every animal will meet those high expectations. They regularly "sort through" their dogs, exchanging animals that are not performing well for others that, they think, have the potential for doing better.

Some of the service-dog organizations raising animals for the disabled, for example, will start five hundred to a thousand well-bred puppies a year—only to have less than 50 percent achieve an appropriate level of working ability. Dogs that fail are not always adoptable as pets for many reasons. They may have been raised in kennels with a large number of other dogs and have not been socialized for the home environment. Some military and police dogs can be dangerous without expert management. And often, working dogs are more active than the average pet owner wants. A retriever may constantly solicit saliva-laden ball throwing; Border collies can hardly be dissuaded from chasing cars, or they go into catatonic trances in the living room while they anticipate that some object, like a ball, might move. Conversely a livestock-guarding dog might show less activity and less "playfulness" than you'd like to see in a family pet. A good one won't ever chase a ball.

In spite of this variability, it is clear that many aspects of the behavior of working-dog breeds are the result of intentional breeding (Darwin's artificial selection) for the expression of particular motor patterns. This sort of guided evolution was surely the case for types of bird dogs used in sport hunting, for instance. A good bird dog is expected to display the FREEZE motor pattern (we'll use capital letters to indicate a specific motor pattern) in the presence of a sitting bird: to stop moving and stand stock-still. The bird dog then displays the POINT motor pattern toward the hunter's target prey: its head, muzzle, and back are oriented on a straight line to the target, and the dog is expected not to move (until commanded). With other types of hunting dog, such as the retrievers and spaniels, breeders rigorously selected against this motor pattern: they didn't want a dog to FREEZE and POINT to a new bird while it was supposed to be accomplishing the desired task of retrieving an already downed bird.

The herding dogs provide a great example of breeds that exhibit some interesting variability in the expression of common motor patterns. The cattle dog, which nips at the heels of cows, was bred to express CHASE (a burst of rapid movement toward prey or other objects of interest) as well as GRAB-BITE (where the animal seizes an object with its jaws). The familiar bulldog, known traditionally as the butcher's dog owing to its use in catching and subduing straying cattle, is another good example of this type, with the added advantage of a particularly effective jaw shape, size, and biting force—and a reluctance to let go. In a sense they freeze on the GRAB-BITE. Many attack-trained military and police dogs are selected for this kind of hypertrophied GRAB-BITE. You also want this behavior in retrievers, which search for a bird that has been shot and then display a form of GRAB-BITE. It's essential, however, that they don't show "hard mouth" and go on to inflict a KILL-BITE, DISSECT, or CONSUME motor pattern. Hunters don't want a dog that crushes the bird and then takes it apart and/or eats it before delivering it to hand. Those motor patterns are considered faults, and dogs are disqualified if they display them. (Shortly we'll see how all these motor-pattern components come together in a complex sequence in predatory carnivores.)

By contrast, sheep-herding dogs like the Border collie were intentionally bred not to express some of these motor patterns. GRAB-BITE is a disqualifying fault if they show it at a trial—and more important, from a working standpoint, a grabbing bite on a sheep might tear and will stress the victim and its reaction may panic the flock. Displaying KILL-BITE, DISSECT, and CONSUME, of course, entirely misses the point of the working Border collie's job. Rather, what you want to see in the Border collie is the expression of a motor-pattern sequence known as EYE > STALK: the dog intently fixes its gaze on a sheep and subsequently (as indicated by ">") slowly moves toward it, and then (on crossing some threshold space) displays the CHASE motor pattern, which causes the flock to be moved in a desired direction.

How do we know CHASE is a motor pattern? Because you can select for or against it, and you can select for dogs that have a specific shape to the chase. Border collies run with a circular motion ending up 180 degrees away from the handler, while the drover's dogs chase cattle directly away from the handler, moving them along a road (fig. 11). Each handler selects dogs that have a specific shape for the specific motor pattern. Border collie breeders have long bred selectively for each of

FIG. 11   THE CHASE PATTERN OF BORDER COLLIES IS CIRCULAR OR "HEADING." BREEDS
LIKE THE DROVING DOGS ARE "HEELERS," WHICH TEND TO RUN DIRECTLY AT THEIR PREY.
WOLVES ALSO TEND TO CIRCLE THEIR PREY WHEN THEY HUNT IN GROUPS. MORE ON THAT
IN CHAPTER 9. MIDDLE PHOTO BY MONTY SLOAN/WOLF PARK.

these motor patterns. Another way of saying this is that dogs that display the proper motor patterns get bred together, and those that display "improper" motor patterns don't.

HOW TO DESCRIBE MOTOR PATTERNS

The first step in doing ethology in the field—whether we're looking at working dogs or wolves—is to put together an inventory of motor-pattern behaviors, informed best guesses about the set of the shapes of behavior that are adaptive in the life of a particular species or breed. Such an inventory is called an ethogram: a compendium of detailed descriptions of bodily shapes and movements that the observer thinks are adaptively significant components of species-typical behavior.

You probably already have a good mental image of what a motor pattern is like. Think of a cheetah, for instance, when it is hunting gazelle. Having detected its prey, the cheetah begins by taking on a stalking shape: it slowly approaches the prey with its head lowered and body close to the ground. At some point, a threshold distance, it changes shape, putting on a running burst of speed toward its prey. If the gazelle can't escape, the cheetah SLAPS it with a forepaw knocking the animal to ground (fig. 12).

The ethogram of motor-pattern expression in the cheetah—ORIENT > EYE > STALK > CHASE > FOREFOOT SLAP, GRAB-BITE > KILL-BITE—is relatively easy to observe even at a distance (and it's the subject of plenty of nature documentaries you can watch in the comfort of your living room) and to describe as a sequence of events in time. Paul Leyhausen, in his famous book *Cat Behavior*, taxonomically classified the wild species of cats by whether they had the FOREFOOT SLAP or not. Some species of cats have a modified form of SLAP, a FOREFOOT CLAP, where the animal leaps into the air and claps its two forepaws to catch an insect, for example. The presence or absence of a particular shape of motor pattern in the ethogram is as much a taxonomic characteristic of a species as the number of teeth it has.

Some motor patterns are, of course, harder to visualize directly. Consider, for example, the small and intricate movements that an animal makes inside its vocal tract—its mouth, throat, and larynx—to produce

FIG. 12 MANY OF THE CAT SPECIES, LIKE CHEETAHS, EXHIBIT THE FOREFOOT-SLAP MOTOR
PATTERN DURING PREDATION. A DOG WOULD HAVE GONE AFTER THE PREY WITH MOUTH
AND TEETH. PHOTO BY DANIEL STEWART.

the sound of a particular species-specific call. Identifying and measuring these sorts of internal movements can be a challenge. Nevertheless, the ethologist's job is to try to characterize each moving "behavioral shape" that we think is functional in the life of an animal and to determine by (usually painstaking) observation how these adaptive patterns are structured and used.

In order to construct an ethogram, to capture the full character of individual motor patterns in the animal's repertoire, we have to precisely describe them on three dimensions: quality, frequency, and sequence. By "quality" we mean the physical characteristics that define the animal's shape as it occupies space at a particular moment in time; "frequency" is a description of how often, over time, a quality state is expressed; and "sequence" is the temporal order in which successive states may occur. Two species might show an identical motor pattern but will differ in how often each species displays that pattern or will differ in where the motor pattern appears in a sequence.

QUALITY: THE SHAPE OF MOVEMENT

The quality of a motor pattern is the overall picture of the movements that we recognize as comprising a particular behavioral shape. Describing the pattern's quality provides us with a systematic basis for putting a convenient unitary name like WALK to what we see an animal doing. Quality isn't a matter of why it's doing something, or how well, or what an animal might be feeling about it. The quality of a behavior is all about its shape, not its function.

Consider the canid motor pattern that is often called an open-mouth GAPE (fig. 13). We can describe its quality thus: the jaw is lowered, slightly exposing the mouth cavity; the skin around the mouth cavity is pulled slightly back from its resting position and the skin on top of the muzzle is wrinkled; the lips are retracted to expose some portion of the upper row of teeth. Anatomists, physiologists, and neurophysiologists might be able to characterize the quality of a motor pattern in an even more fine-grained way by investigating and describing "lower-level" anatomical events: precisely which muscles move, what the exact sequence of muscle movement is, which neural control mechanisms are engaged, and so forth. Usually that isn't feasible—certainly not for working field ethologists—so we content ourselves with describing and labeling the overall shape of the animal's movement that results from the action of these underlying physical mechanisms.

A canid motor pattern like the GAPE is often identified as a defensive aggressive behavior in dogs, a self-defense (hazard-avoidance) mecha-

FIG. 13   THE GAPE MOTOR PATTERN IS SEEN IN THE WOLF (*PICTURED HERE*) AND THE
COYOTE, BUT ONLY RARELY IN DOGS. IN FACT, SOME OF US HAVE ONLY EVER OBSERVED
IT IN YOUNG JUVENILE DOGS. PHOTO BY MONTY SLOAN/WOLF PARK.

nism. People who know dogs well are likely to be wary of an animal that
expresses it. However, when we construct an ethogram we need to be
agnostic about the purpose (let alone the emotional content or cognitive experience) of a motor pattern. It is possible that the animal could
be using the GAPE as part of play, as we will see in chapter 9. We must
leave open the question of its functionality until we can analyze the outcomes of a behavior and the contexts in which it occurs.

Consider the smile or grin that we see in both human beings and
chimpanzees—this might (or might not) be related to the somewhat
similarly shaped GAPE in other mammals. We observe smiling in all
members of these species. Even very young humans and chimps smile,
and smiling is universally recognized across human cultures. So we
might say that each species exhibits an intrinsic motor pattern that engages essentially the same oro-facial muscle movements—and so appears to have roughly the same quality in both animals.

Consequently, an ethogram of the each species would include a very similar description of this motor pattern. We can call it anything we like. SMILE will do, and we could choose to use the same name in the ethograms of both humans and chimps. But we always have to remember that this is an informal label for the motor pattern's quality. The fact that the two animals may seem to share almost-identical motor patterns doesn't necessarily mean that they share the same motivational or emotional states or that the motor pattern has the same adaptive function in both species. The primate SMILE is a case in point. When we look more closely at the contexts in which SMILE occurs in many nonhuman primates, we find that this motor pattern (like the open-mouth GAPE in dogs and other canids) seems to be associated with fear and, possibly, ensuing aggression. This is not, however—at least not obviously—the role that smiling plays in us humans. (It might, however, be a kind of low-level threat that is associated with fear; we knew a botanist at our college who only seemed to smile when he was angry and annoyed!) We and the chimp may move in similar ways, but what our actions mean behaviorally might be another story entirely.

Another example of shared quality can be seen in the FOREFOOT-STAB motor pattern of some predators, where the animal leaps into the air, extends its forelegs and then lands, forefeet first, on small moving objects, such as prey animals like mice (plate 3). We see it in wolves, coyotes, foxes, border collies, some other breeds of dogs (but not all), and in some species of cat. It is, for all intents and purposes, identical in all of them.

The main "take-away message" is that a species-typical motor pattern, as Lorenz and Tinbergen said, is a taxonomic characteristic—and a product of the genes. When several species perform a nearly identical motor pattern in the same sequence, it is likely they have inherited the behavior from some common ancestor. Similar traits seen in different but related species are called homologies (though they still should be regarded as species-typical behaviors within any one species). In figure 14 a wolf and in plate 3 a dingo are performing a FOREFOOT-STAB, a component of predatory behavior (which we will discuss in greater detail in the following chapter). There is little doubt that we are looking at an identical—homologous—motor pattern. Homologous behaviors, like physical traits, can be used to categorize family relation-

FIG. 14 THE FOREFOOT-STAB. WOLVES, COYOTES, FOXES, ALL THE CATS, AND MANY
OTHER SPECIES OF CARNIVORES ALL PERFORM THIS MOTOR PATTERN, SUGGESTING THAT
IT IS VERY OLD FROM AN EVOLUTIONARY STANDPOINT—PERHAPS FOUR OR FIVE MILLION
YEARS. PHOTO BY MONTY SLOAN/WOLF PARK

ships among species and to illuminate their evolutionary histories. The
forelimbs of mammals and the wings of birds are famous examples of
structural homology. Physical homologies have commonly been used
when addressing taxonomic issues, but by exploring shared behavioral
patterns, ethology offers additional tools for understanding speciation
and evolution, two of the biggest questions in biology.

We believe, for example, that the modern canids and felids, dog and
catlike animals that all display FOREFOOT-STAB, did indeed descend
from a common ancestral group, the Miacidae, over forty million years
ago. It's true that we only know about the ancient miacid mammals
from fossil bones; behavior rarely leaves a fossil record. However, since
FOREFOOT-STAB is homologous—has the same quality—in diverse
but clearly related contemporary species, it's plausible to hypothesize
that this particular motor pattern was also a behavioral trait of their
most recent common ancestor (albeit an ancestral species that may be

very distant in time). Thus we can imagine, at least, that a long-extinct miacid carnivore—now observable only as a skeletal museum exhibit—might once have leapt into the Eocene air and stabbed at small prey with its forepaws.

FREQUENCY: HOW OFTEN AN ANIMAL DOES IT

Quality is the description of an animal's state in space; frequency is the parameter that describes the occurrence of that state over time. A particular species may exhibit a behavior regularly and quite often, or sporadically, or only very rarely. Barking, for instance (assuming for the moment that it really is a distinct single motor pattern—see our discussion in chap. 8), is only infrequently observed in wolves whereas it is ubiquitous among most dogs, so frequent that it is often described as a hypertrophied behavior. Greyhounds in Argentine villages, in contrast, rarely bark at strangers, but Maremmas in Italian villages rarely refrain from it. Elephant seals walk infrequently, and then only during courtship performances in the springtime when they move on the beach. If you observe them only in winter or summer you may never see them walking.

Behaviors can also change in frequency over an animal's lifetime. Wolves, in fact, exhibit barking more frequently as young adolescents, but as we've said it is rare in adults, except when they are threatened at a den site—and then the frequency goes up. Dog pups bark first at seven days of age and then usually persist in very frequent barking throughout their lives. Thus, one of the major differences among wolves and coyotes and dogs and the various breeds of dogs is the frequency at which they display BARK motor patterns.

The environment certainly plays a role as well: prey size, for example, can change the frequency of FOREFOOT-STAB and HEAD-SHAKE in predators. Likewise, kenneled Border collies tend to display an annoyingly high frequency of barking but display the motor pattern far less frequently when they are working with sheep in the field. It is instructive to note that we had some congenitally and profoundly deaf Border collies who couldn't hear barking in other animals but barked like other collies, both in the quality of the call and its frequency. The behavior clearly wasn't learned. This suggests that the frequency with which a motor pattern can be displayed may, like its quality, be a taxonomic character of a species or breed.

SEQUENCE: THE ORDER OF MOVEMENTS

When you watch a predator hunt, whether it's a wolf, a lion, or a leopard, the process can seem graceful and seamless. The observant ethologist, though, sees a more complex chain of events. In fact, these predators engage a set of discrete motor patterns that are deployed as an integrated sequence of changes in shape over time.

A wolf, for instance, having detected and then oriented toward a prey animal, first goes into the EYE motor pattern: it stands or lies stockstill, staring fixedly at the prey (plate 2). Then it moves into STALK: body lowered, head down but holding the prey in sight, the wolf moves slowly forward. The next step in the chain is the CHASE: the animal shifts into a full-speed forward running movement. When it reaches its target (if the prey object hasn't escaped), the wolf makes an immobilizing GRAB-BITE with its mouth: jaws and teeth closing over the prey animal's leg or haunch. The GRAB-BITE can sometimes tear tissue and bleed the prey to death, and large predatory cats can suffocate their prey with a GRAB-BITE to the neck. Not infrequently, a more invasive KILL-BITE then ensues, typically to the neck, opening the carotid artery or jugular vein. When the target is down, the wolf uses its teeth to DISSECT tissue and expose internal organs. (DISSECT motor patterns are so species typical that a good field biologist can look at a carcass and tell which predator species killed it.) Then the wolf moves on to CONSUME the prey. Each of these behavioral shapes is a distinct motor pattern with its own entry in the ethogram, the behavioral "rule book" of the species. Each can occur independently but also in sequences over time and in combination with other components of behavior.

COLLECTING MOTOR-PATTERN DATA

Once the ethologist has constructed an ethogram, the hard work begins — collecting data "on the ground" about what animals are actually doing. A relatively easy way to get a general overview is by ad libitum sampling. An animal is always doing something (even if only quietly resting or sleeping), and when it is active, a great many things may be happening, sometimes simultaneously. In ad libitum sampling, the ethologist isn't driven by a special interest in a particular type of behavior or a specific hypothesis—the idea is just to get a wide-angle picture of what is going on, to get a handle on what seems to be behaviorally significant in an essentially continuous stream of movement. You may be watching a single animal or a group or shift your attention from one to the other. What you observe and note down is likely to be pretty random, so ad libitum sampling doesn't necessarily yield reliable scientific data. But it can provide a pretty good seat-of-the-pants sense of how a particular kind of animal behaves, as well as a road map for future research.

Focal animal sampling is a more systematic approach. In this method, the ethologist focuses, eagle-eyed, on a single individual, and every instance of that individual's motor patterns is noted and recorded during some predetermined time frame. All-occurrence sampling is another way to do it. Instead of focusing on a single individual, the researcher hones in on a particular motor pattern: how many times is CHASE behavior observed (no matter who is doing it)? If we're interested in social interactions in a group of animals, this can be an especially fruitful approach.

Whichever observational method is employed, it basically comes down to noting that an instance of a particular behavior has happened. Don't forget, however, that the temporal dimension—how long a motor pattern lasts, how often it occurs, and how regularly it may be repeated—is critical as well. In behavior, remember, "time is money." Action requires energy, and the longer a behavior lasts and the more often it is expressed, the more calories the animal needs to obtain to sustain the movement.

But actually measuring the frequency of occurrence and the time course of motor patterns is a tricky business. We want to know, for instance, how long a predatory CHASE lasts and how many times it is repeated during a day. Ideally, an ethologist would have to observe an animal or group of animals continuously over long stretches of time, attempting to record every instance of a particular pattern and its duration. That's usually quite impractical. Even ethologists need to stop for a bit of food and sleep. Video recorders don't have that kind of problem and can be set out to acquire data automatically for long continuous periods. One virtue of the video recorder is that more than one observer can look at the action and so ensure that there is interobserver reliability, or consistent interpretation of the data. A second great advantage is that video data can be stored and looked at again (and again) to provide a more fine-grained description or to test new hypotheses. The problem with unattended recording, however, is that the field of view of a camera is limited. Human observers can quickly take account of events happening behind them, for instance, or shift attention to the edge of their visual field. That said, both direct human observation and the careful analysis of recorded video data are now often employed in ethological studies.

A second general problem is that we need to be sure that the motoric events we observe in sequence are in fact part of the same (functional) pattern. Suppose an animal exhibits a motor event, A, that is usually followed immediately by a second event, B. It's reasonable, by hypothesis, to presume that they are components of a motor-pattern sequence. But what if B occurs two minutes after A? Are these separate events, or is this an instance of the delayed expression of a full sequence? We can't always be sure. Nevertheless, we do need some way of systematically acquiring and describing data. Time-sampling techniques have been devised to try to impose some methodological consistency.

There is a variety of such methods, usually applied in the context of focal animal sampling. One is the so-called 1/0 technique, whereby an observer scores a behavior as either occurring (1) or not (0) during a specified time period. A second approach is sometimes referred to as the instantaneous method, whereby observations are made at regularly specified points in time, for example, every minute on the minute, or every hour on the hour. There's a lot of debate about which time-sampling methods are better, which most adequately capture the full picture of temporal sequences as well as the actual frequency of motor patterns. Each method has its use and value in measuring particular types of behavior in particular contexts. And whatever method is employed, it's important for more than one observer to analyze the same data in order to insure that the description doesn't reflect biases (if unintentional) that the primary researcher may bring to his or her analysis.

It can be a burdensome challenge to meet these methodological requirements, and worrying about them may perhaps seem dry as dust. But what we conclude about the behavior of animals can't be based either on easy casual observations about what "my dog" does or on appealing stories in the popular media. Good science needs systematically collected data acquired by reliable and consistent methods.

MOTOR PATTERNS EXPRESS RULES

When we use a term like "pattern," we are talking about nonrandom activity. Randomness essentially means unpredictability. As scientists, we want to be able to predict things—that's how we test the validity of a hypothesis—and if behavior were random we wouldn't be able to say much (if anything) about what is likely to happen in the life of an animal. It is undeniable, of course, that some motor events are contingent on essentially random events or states of the environment. A running animal that suddenly encounters a patch of ice may slip and slide, careening unpredictably. The motor-pattern sequences that concern ethologists, however, are not random responses to accidental contingencies or quirky individual movements. They are systematic responses to specific internal motivations and external stimuli and signals. Lorenz and Tinbergen called these triggers of motor-pattern "releasers." The detection of a prey animal releases predatory foraging behaviors in a carnivore. The appearance of a predator may trigger vocal alarm-calling behavior of the prey directed at other group members; the alarm signal releases flight or defensive behavior. In a complex motor-pattern sequence in a single individual, one motor pattern may be motivated by the prior occurrence of another. John Fentress and P. J. McLeod very nicely captured these dependencies when they wrote that "it is through an integrated sequence of movement that an animal expresses rules."

We normally describe and measure motor patterns as moving shapes of the animal. As Fentress and McLeod suggest, we can also think of them as sets of rules (like an algorithm or a computer program) that govern those shapes. Rules have the general form "under condition X, or if Y occurs, do Z" (display *this* motor pattern). The notion of "rules" can sometimes be a bit confounding since in everyday life we tend to understand a rule to be an explicit instruction or restriction imposed by some authority. Another way to think about rules is that they are simply descriptions of a nonrandom action or structure; they set limits on the state and activity of those properties. In a game of checkers, for instance, rules determine how and in what direction the pieces can move and what moves are possible in a particular board configuration, and they set the order and frequency with which the moves can occur.

Think about a mundane human behavior like ordinary walking, for instance—a perfectly good example of an animal motor pattern. There is a simple nonrandom rule-governed "recipe" for this common activity:

- From a standing position, raise one leg to a certain angle while flexing the knee and begin to fall forward.

- Move that leg forward and put that foot down, stopping the fall.

- Unflex the knee, bringing the body up to full height.

- Then the other leg and foot.

- Repeat.

These motor actions embody an ordered sequence of rules for the behavior of walking.

Normally, of course, we're completely unconscious of these rules. They are rarely if ever made explicit to us in normal life. Indeed, if you try to think about and follow them consciously and then try to walk, your movement is likely to be jerky and awkward. When you follow the rules unconsciously, the outcome is generally smooth locomotion. If you do try to bypass the rules of the "walking game" and intentionally violate them, say, by failing to apply the knee-flexing rule or by randomly carrying out the component actions that are involved in walking, the result will be bizarre. Think about goose-stepping soldiers—or John Cleese's "silly walk" in his famous Monty Python sketches (fig. 15).

This sort of movement looks odd (and funny) to us because it transgresses the rules that define the species-typical adaptive pattern. In-

FIG. 15 THE "SILLY WALK" POPULARIZED BY THE MONTY PYTHON COMEDY GROUP. YOU
CAN LEARN TO WALK THIS WAY BUT IT IS NOT A SPECIES-TYPICAL MOTOR PATTERN. IT IS
METABOLICALLY VERY "EXPENSIVE," AND YOU CAN ONLY DO IT WITH SURPLUS ENERGY
RESOURCES—AND A LOT OF COMIC INGENUITY. DRAWING BY CAROL GOMEZ FEINSTEIN.

deed, trying to sidestep the normal walking rules (pun intended) is
energetically expensive—and it can be dangerous to your health. So it's
a good thing that we don't need to learn to walk "correctly." People often
say that young children learn to walk, but all the evidence suggests that
it is a behavior that appears automatically in development and doesn't
require either an adult model or explicit training. We don't—we can't,
in fact—teach our babies to do it. A baby will begin to walk on its own.
Anxious parents whose children are late walkers know that you can't
hurry the process along. Nor can we train children, or even ourselves
as adults (unless we are regimented soldiers or consummate comedic
actors), to regularly walk in an odd or random fashion by contraven-
ing the adaptive rules of walking. Exactly the same thing is true of the
motor patterns that any animal exhibits. Dogs don't learn to BARK;

a Border collie doesn't have to be trained to display EYE > STALK or CHASE; a wolf doesn't need to learn the KILL-BITE.

So the behavioral shape we humans display when we express the WALK motor pattern, when we "follow the rules," is an intrinsic species-specific property of the human machine. That is not to say that individuals won't vary a bit (or even quite a bit, sometimes) in the precise fashion that each one implements the species-typical rules. Some people walk with their feet slightly more splayed or turned inward than average, or at a slightly faster or slower rate. Often we can identify people we know at a distance by these small deviations from the species norm. When we look at motor patterns in any species or any population we will find this sort of variability. Indeed, it is variation that makes natural selection and improved fitness of the species possible.

Individual development is, of course, critical as well. Walking, like other motor patterns, is subject to timing and growth factors. Newborns don't do it: normal walking depends on the organism attaining juvenile/adult skeletal proportions and neuromuscular controls. We also have to take into account environmental effects and the ordinary contingencies of everyday life. The usual WALK motor pattern can be disrupted if you inadvertently have a stone inside your shoe: you're likely to change your shape—alter your gait and weight distribution—in order to minimize discomfort. You can often tell, simply by observing shape and rule-implementation differences, that someone is in this uncomfortable situation. Likewise, a dog with an amputated leg may be able to trot along but it won't display the quality of the normal behavioral shape that we expect to see in a trotting dog.

Finally, you can also think of the description of a motor pattern as an indirect or abstract reflection of species-specific brain shapes in the behaving animal. In all complex organisms, it is the central nervous system that actually determines and implements a movement. Intrinsic behavioral rules are implicit instructions to engage in actions that must be initiated by particular patterns of connected nervous tissue. If you are "wired" like a duck, you will (all things being equal) walk like a duck and quack like a duck.

This relationship between brain shape and motor pattern is illustrated rather remarkably by the work of a group of French neuroscientists at the National Scientific Research Center and College. They took tissue from the brains of Japanese quail chick embryos that is believed to regulate how members of that species make a particular call and transplanted it into the embryos of domestic chickens. Ten days after hatching, the chickens produced the distinctive notes of a quail's call. If members of a species are wired the same way—if their genes build similar brains—they will be disposed to behave in a species-specific fashion, to exhibit species-typical motor patterns. Change the shape of the nervous connections in the machine and you change its behavior.

5 THE RULES OF FORAGING

Foraging—finding, acquiring, and consuming food—is an especially good way to look more closely at the story of motor patterns and sequences. It's very easy to observe motor patterns that support foraging since animals always have to feed, virtually continuously, in order to acquire the energy they need throughout life. With domesticated animals like the household dog, we humans provide the food and we're always right on hand to see what transpires. Hazard-avoidance behaviors, in contrast, are less frequent and thus more difficult to observe—one might imagine that an animal "hopes" never to have to resort to them! And reproductive activity is often seasonal, only available for observation as rarely as once a year in many canid species.

We'll start by focusing on the foraging motor patterns of wild canid carnivores, like the wolf and coyote, and big cats, like pumas, for whom meat from freshly killed animal prey is a typical primary food resource. What defines a mammal as a member of the order of Carnivora is its dental shape: all carnivores have specialized teeth called carnassials, which can be used to cut and tear animal tissue. Dogs are carnivores, too. But when we say that an animal is a carnivore it doesn't necessarily mean that it always and only eats meat, let alone meat that it has captured itself. Some members of the order Carnivora are indeed obligatory meat eaters. There are cat species, for instance, that require a dietary enzyme that they can only get from fresh meat. By contrast, the giant panda is taxonomically a carnivore but in fact it is obligatorily herbivorous, specializing in bamboo leaves. Other members of the genus *Canis* are great scavengers and opportunistic feeders. In reality, while wolves do hunt and kill large prey animals like moose, they also consume prey that was recently killed by others, and they can supplement their diet with many nonmeat foods as well.

We and our students once carried out a study on coyote foraging in a large forested reservoir area with a significant population of white-tailed deer and found that, in every month of the year, berries or other fruit were still the largest component of the diet of these "carnivores." Many of our own observations of wolf feeding behavior were made by waiting for them to show up at garbage dumps! The dog itself, though so closely related to wolves, rarely if ever hunts for meat. Instead, millions of them consume human-provided commercial dog food, which certainly doesn't look or behave like a moose and sometimes has very little if any meat in it at all. And as every dog owner knows, a dog is usually happy (for better or worse, from a nutritional standpoint) to scarf up just about any kind of scrap from the table.

Nevertheless, when we look at the big picture of foraging behavior in the carnivores, we observe a common set of motor patterns that are typically associated with predation: hunting, capturing, killing, and eating prey. Here is the typical predatory motor-pattern sequence:

ORIENT > EYE > STALK > CHASE > GRAB-BITE > KILL-BITE > DISSECT > CONSUME

This schema consists of a series of discrete motor components normally occurring in a particular order (the ">" symbol means "precedes"). It describes an ideal chain of events, the rule book for the form of foraging behavior that carnivore predators deploy as an integrated sequence of changes in shape over time and space.

A wolf, having detected and then oriented toward a prey animal, first goes into the EYE motor pattern: it stands or lies stock-still, staring fixedly at the prey (fig. 16). At least, that's what we visually oriented human ethologists tend to assume—that vision is the key input here, providing the releaser that triggers the motor pattern. If you watch carefully a wolf, coyote, or Border collie will also have its mouth dropped open during what we call the EYE motor pattern. In fact, like many

FIG. 16 THE EYE MOTOR PATTERN. THE HEAD IS MOTIONLESS AND OFTEN LOWERED. THIS
WILL BE FOLLOWED BY STALK, CHASE . . . AND ULTIMATELY THE KILL-BITE. THE SEQUENCE
IS ALMOST IDENTICAL IN MOST CARNIVORES. PHOTO BY MONTY SLOAN/WOLF PARK.

mammals, they may be sensing the prey with a specialized vomeronasal organ, a highly vascularized structure that sits on top of their palate at the base of the nasal cavity. This organ is exquisitely sensitive to certain chemical cues that may be picked up when the mouth is ajar. So when we informally say that a carnivore is looking at the prey—in the EYE motor pattern—it may be that they are also "looking" with their organs of smell.

The predator then moves into STALK, with its body lowered and its head down, directed toward the prey. (See plate 2.) Holding the prey in "sight," the animal moves slowly forward. The next step in the chain is the CHASE: its shifts into a full-speed forward running movement. Two kinds of BITE behavior may then ensue: a grabbing bite that effectively disables the prey, and a killing bite that dispatches it. The carnivore then uses its carnassials and other teeth to DISSECT its victim. If all goes well, the sequence ends when predators CONSUME their prey.

The bet is that natural selection has been at work in the evolution of these particular behavioral shapes in carnivore predators. If the full sequence is engaged, if each motor event occurs in the right order (and if the target prey animal doesn't manage to escape), the behavior tends to result in an adaptive functional outcome—the predator acquires food energy that fuels its machinery. Selection, however, doesn't necessarily provide a single optimal solution to every problem, a best shape or single rule that fits all needs for all animals. Thus we find that different motor patterns are sometimes substituted in the sequence. Recall the FOREFOOT-STAB we illustrated in the previous chapter. This motor pattern expresses an "optional" rule and is released by small prey like mice (or, more accurately, by small moving objects that are detected in the environment) rather than large prey animals like moose or elk. When this rule is engaged, then, wolves and coyotes (and some dogs) may substitute (plate 3) the FOREFOOT-STAB for CHASE as well as a HEAD-SHAKE for the KILL-BITE:

ORIENT > EYE > STALK > FOREFOOT-STAB > GRAB-BITE > HEAD-SHAKE > DISSECT > CONSUME

This alternative pattern is clearly adaptive in the context of a different type of prey. Large cats like the puma (mountain lion) make substitutions as well, engaging different strategies in different circumstances. When hunting deer, these powerful predators with large clawed paws sometimes replace the GRAB-BITE with FOREFOOT-SLAP and then go on to inflict a KILL-BITE. Pumas may also lie in wait and then POUNCE on a deer from a hiding spot. Thus we see:

ORIENT > EYE > STALK > CHASE > FOREFOOT-SLAP> KILL-BITE > DISSECT > CONSUME

Or:

ORIENT > EYE > POUNCE > KILL-BITE > DISSECT > CONSUME

Interestingly, domestic dogs differ from their wild carnivore relatives in a fascinating way: they rarely engage the entire (functional) predatory sequence. Moreover, different breeds exhibit quite distinct partial sequences. We'll look more closely at these breed differences shortly and try to puzzle out why the full behavior is atypical of dogs. The general story, however, is that, over the course of evolution, different prey types and foraging conditions have given rise to modifications in the general carnivore pattern. This is just what we should expect if behavior is indeed an adaptive product of evolution that is subject to specific natural selection pressures in different niches. But as we've also seen, certain commonalities have long persisted in the predatory motor-pattern sequences of all the carnivores, reflecting their shared ancestry (phylogeny). Successive generations of carnivores have all inherited an intrinsic "wired-in" program of rules that at least partly governs their predatory behavior.

The physical shape of different carnivores plays a significant role as well. The wolf and the puma, for instance, both have a form of KILL-BITE — but the actual quality of this general motor-pattern component is slightly different in each species. The puma uses KILL-BITE to suffocate its victims either by putting a hard bite on the throat, collapsing the trachea, or by actually grabbing the muzzle so the animal can't breathe. Wolves tend to GRAB-BITE the hind legs and tear them open, killing the prey by slowly bleeding it to death. Pumas and wolves have differently structured jaws, dentition, and musculature and can apply different mechanical forces. As puma and wolf shapes differ, so their behavior differs.

Each of these behavioral shapes, each distinct motor pattern element, has its own entry in the ethogram, the behavioral rule book of the species — and each can occur independently or in combination with other components of behavior. In order to successfully forage, to dissect and feed on prey, a puma has to go through the whole sequence — EYE > STALK > CHASE > GRAB-BITE > KILL-BITE with, perhaps, a substitution. But dogs, wolves, coyotes, and jackals can enter the sequence anywhere. Canids can pass over the initial motor patterns and start with DISSECT and CONSUME—they can scavenge already dead animals. You'll recall that, back in the twentieth century, when the U.S. government decided to get rid of livestock predators by spreading poison baits on the range, the system worked well on wolves and coyotes, species that were totally eradicated in some regions. But it didn't work well on pumas: they aren't "taken in" by poisoned dead bait because they have to go through the whole foraging sequence in order to finally get to DISSECT and CONSUME. Another way to put it, as we suggested earlier, is that the motivation for dissecting, its releaser, is the KILL-BITE itself; the motivation for the KILL-BITE is the GRAB-BITE; and so on. So pumas won't—in a sense they actually can't—eat already dead animals. Indeed, in another study we did in Namibia, we learned that newborn calves had a better survival rate during a cheetah attack than older cattle simply because the newborns did not run. Without the releasing trigger of a running prey animal, the cheetah wasn't motivated to CHASE or to engage any subsequent motor patterns necessary to achieve a kill. (And once again this brings into question whether these predators have any sort of intelligent grasp of their goals or, indeed, whether they are conscious of their own actions.)

As noted, wolves can enter the sequence at any point. They can, for instance, simply start the foraging pattern at DISSECT: that motor component doesn't have to be released by the prior performance of KILL-BITE. Alternatively, if wolves are already engaged in the EYE > STALK > CHASE sequence and get interrupted (perhaps by a noisy livestock-guarding dog!), they often cannot proceed. It appears that CHASE is a releaser for GRAB-BITE and any delay after a halted (or aborted) chase diminishes its effectiveness as a trigger for the next steps.

GRAB-BITE, therefore, almost always appears after CHASE and before KILL-BITE. It also occurs between CHASE and the HEAD-SHAKE substitution in the sequence. Consequently, GRAB-BITE is a (potential) releaser for both HEAD-SHAKE and KILL-BITE, in wolves and coyotes alike. But since HEAD-SHAKE itself occurs much more frequently in coyotes than in wolves, we should also expect to observe a higher overall frequency of the triggering GRAB-BITE in coyotes.

Is this difference in frequency an intrinsic genetically determined property? Are wolves and coyotes "wired up" to express these motor patterns at different rates? Well, we can't tell easily because we can't control their environment or the size of the prey that they pursue. HEAD-SHAKE is, in fact, elicited by smaller prey, and coyotes tend on average to hunt smaller prey. If the two species shared the same geographic area with equal access to all prey, would coyotes specialize in smaller prey animals and wolves in the larger? The answer may well be yes. Coyotes may simply prefer smaller prey: properties of their intrinsic shape—size, speed, or even visual acuity—might bias their prey choice. These factors would yield different observed behavioral frequencies even if the two animals do, in fact, share an identical intrinsic set of motor patterns. Coyotes also may face an additional problem. If they pursue the same prey as wolves, they will be vulnerable to attack by those larger and stronger competitors. Before wolves were reintroduced in Yellowstone National Park, groups of coyotes were observed to hunt and kill big game, but following wolf reintroduction, this behavior all but ceased.

PATTERNS DEVELOP

There's another dimension to this story—the role of development. When talking about wolf predatory behavior, most people tend to think in terms of what adult wolves are doing. What often gets forgotten is that during each stage in a developing life an animal has its own unique shape, and there are unique foraging rules that follow from it. The fact is that we normally think about animals in terms of the adult alone—as if the adult form were the goal of growth. "Why don't you grow up and act like an adult?" our mothers say to us. Our image of "the wolf" is the adult shape. We don't think about newborn and young wolf puppies (which look almost indistinguishable from dog pups).

But puppies—wolf or dog—are just not the same organisms as adults. Their shapes are very different, and puppies are rapidly changing over time. Puppies aren't just simpler and smaller versions of the adult: they have their own quite complicated behaviors (and different energy requirements). Indeed, some aspects of their physical form may be considerably more complex than the adult's. Puppies, for instance, have much finer muscular control of the mouth and tongue than adults do. This is likely a special adaptation to the way that newborn mammals feed, by suckling from the mom's teat. This kind of oral motor control diminishes and disappears as the animals grow. Similarly, as we'll see in a subsequent chapter, newborn pups have an intrinsic care-soliciting vocal signal that disappears in mature animals. So it's important not to think of pups (or any young animal) as immature, as an organism that "needs to grow up." No, puppy shapes and the behaviors they support are adaptive characters in themselves.

As a case in point, let's look at wolves as they progress through three sets of foraging motor patterns during their lives—and compare this to the developmental picture in dogs. When wolf and dog pups are born they immediately exhibit a characteristic neonatal sequence:

ORIENTATION > LOCOMOTION > ATTACHMENT > FOREFOOT-
TREAD > SUCK

In narrative terms, the pup turns toward mom, moves to her, latches on to her teat, moves its front paws rhythmically (physically stimulating the flow of milk in the mom's breast), and draws the milk down by mouth muscle movements. This "foraging sequence" has its onset just about at the moment of birth. It is an archetypal intrinsic behavioral pattern: it requires no experience and is universal in the young of the species. Indeed, an identical pattern occurs in a wide range of mammals. What a neat system! The mother makes a special food, milk, which is easily digestible and "automatically" obtainable by neonatal behavioral rules: pups get their food by engaging a complicated sequence of motor patterns that trigger the mother's rules for presenting it.

But four weeks or so later a pup's digestive system begins to mature—it changes its physiological shape—and now it can switch from milk to solid food. Its swallowing mechanisms change as well; presenting solid food too early will lead to choking in very young pups. A new integrated set of motions that express rules for feeding appears:

ORIENT TOWARD MOTHER'S HEAD > APPROACH > NUZZLE

PLATE 1 THERE ARE SEVERAL MOTIVATIONS FOR DOGS TO BE IN CEMETERIES. IT CAN BE A QUIET AND UNTHREATENING SPACE; THE STONES MAY PROVIDE A WARM PLACE TO LIE; AND BODILY REMAINS MIGHT OFFER UP THE SMELL OF FOOD. IN MANY PARTS OF THE WORLD, DOGS DO DIG UP RECENTLY BURIED BODIES AND FEED ON THEM. PHOTOS BY JANE BRACKMAN (*TOP AND BOTTOM, LEFT*) AND EVIE JOHNSTONE (*BOTTOM, RIGHT*)

PLATE 2 EYE > STALK IS A COMMON MOTOR PATTERN IN THE CARNIVORES. PHOTOS BY CHRISTIAN MUÑOZ-DONOSO (*TOP*), LORNA COPPINGER (*CENTER*), AND MONTY SLOAN/ WOLF PARK (*BOTTOM*).

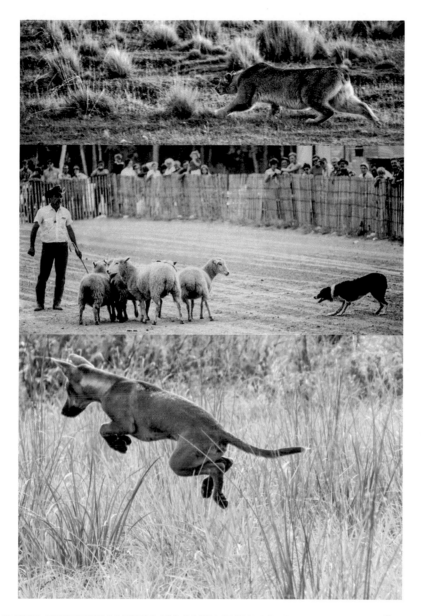

PLATE 3 THREE MOTOR PATTERS: *TOP*, A PUMA IN STALK (PHOTO BY CHRISTIAN MUÑOZ-DONOSO); *MIDDLE*, A BORDER COLLIE SHOWING EYE (PHOTO BY LORNA COPPINGER); AND *BOTTOM*, A DINGO EXHIBITING THE FOREFOOT STAB (PHOTO BY KERRIE GOODCHILD, DINGO DISCOVERY CENTRE).

PLATE 4   PARENTAL BEHAVIOR 1. ALL MAMMALS SUCKLE THEIR OFFSPRING, AND THE CANIDS AND MOST CARNIVORES CONSTRUCT A DEN TO PROTECT THEM. PHOTOS BY BRADLEY SMITH (*TOP*) AND DANIEL STEWART (*BOTTOM*).

PLATE 5   PARENTAL BEHAVIOR 2. WOLVES, COYOTES, JACKALS, AND DINGOES ALL CARE
FOR AND FEED PUPS LONG AFTER THE NURSING PERIOD. DOGS ARE THE EXCEPTION IN
THE GENUS *CANIS*: PUPS ARE ON THEIR OWN AFTER WEANING. PHOTOS BY BRADLEY
SMITH (TOP) AND MONTY SLOAN, WOLF PARK (*BOTTOM*).

PLATE 6   ANIMAL PLAY IS THE MYSTERY OF MYSTERIES. IT CERTAINLY SEEMS TO PEOPLE
THAT THE BEHAVIOR OF THE DOGS, WOLVES, AND DINGOES ON THIS PAGE AND THE
NEXT HAS A PLAYFUL ASPECT. BUT WHY DO WE THINK SO? WHAT IS ITS FUNCTION IN
THE ANIMAL'S LIFE? HOW DO WE KNOW THAT THIS IS "PLAY" AND NOT SOMETHING ELSE?
PHOTOS BY MONTY SLOAN/WOLF PARK.

PLATE 7   MORE ANIMALS SEEMINGLY AT PLAY. PHOTOS BY CHRISTIAN MUÑOZ-DONOSO (*TOP*) AND BRADLEY SMITH (*BOTTOM*).

PLATE 8   HUMAN SOCIAL CUES CAN BE READILY INTERPRETED BY MANY, MANY SPECIES, NOT JUST DOMESTIC DOGS. PHOTOS BY LORNA COPPINGER.

FIG. 17 WOLF PUPS SOLICITING FOR MOM TO REGURGITATE DINNER. PHOTO BY MONTY SLOAN/WOLF PARK.

At this point, young pups aren't entirely adept at chewing adult food. Their teeth aren't full-sized yet and their muscles for tearing and chewing food are just getting into proper shape. In fact, solid food is not yet fully digestible: the pup's stomach enzymes are only just converting from being able to deal with milk to dealing with solids. In addition, a mother is beginning to tire of the process. She isn't producing as much milk as it takes considerable energy to manufacture it, and her teats have been torn at for several weeks. The new food-begging motor-pattern sequence is an adaptive solution to these problems: it stimulates the wolf mother (and perhaps another adult "helper" pack member as well) to regurgitate some of her own food to the pup, who then can easily consume it (fig. 17).

As wolf pups develop their digestive systems, they also use this motor pattern to beg for solid food from adult males who have been off on a hunt and bring back portions of killed prey to moms and offspring at a rendezvous point. This provisioning process precedes the young wolf's full predatory behavior. The system doesn't work this way in dogs, however. Dog pups have the begging motor pattern in common with wolves—but mothers rarely if ever respond by regurgitating, and other adults (including fathers) play no role in feeding the offspring. Fortunately for dogs, who persist in using the motor pattern throughout life, their human caretakers are ready to respond and provide them with food.

These changes are clear examples of the intrinsic character of adaptive motor patterns. They unfold at different time points in the animal's development but are nevertheless "genetic" in nature. Imagine if a newborn pup were required to learn how to beg or to find the teat, how to attach to it, and how to suck. Experimenting with different methods of suckling and different things to suck on would be energetically very inefficient, given the tiny size of the pup, its limited resources, its limited sensory and locomotor systems, and the limited time available to it to learn such a very complex behavioral sequence. It might well be fatal. Happily for the infant, the expression of neonatal foraging motor patterns is a part of the shape of the infant "machine." They are adaptive products of natural selection. You can't teach a newborn pup not to produce the intrinsic suckling motor pattern, and you can't train a pup to forage on its own for solid food or induce it to eat solid food simply by manipulating its behavior in the lab.

Neonatal suckling doesn't simply turn into adult feeding because of the animal's experience or general patterns of growth. In fact, one system is replaced by another. W. G. Hall and C. L. Williams, who studied neonatal foraging behavior in rats (and their conclusions are likely to be generalizable to all the mammals), showed that the brain mechanism that controls infant feeding—intrinsically "wired" with rules for the neonatal sucking motor pattern—is not the same part of the brain that controls the adult foraging motor pattern. These behavioral patterns arise from distinct shapes of the biological machine, separate neurological systems that we think must have evolved independently of one another. Indeed, the mammalian neonatal foraging pattern is a novelty among all the vertebrates. By contrast, the adult carnivore EYE > STALK ... pattern is evolutionarily very conservative and is homologous to predatory foraging patterns that we see in reptiles and even in the more ancient fish.

DOG FORAGING RULES

Domestic dogs share a deep evolutionary history with the other carnivores. Remarkably, however, the modern dog breeds (that is, artificially selected varieties of a single species) typically exhibit only parts of the predatory motor-pattern sequence of their wild canid relatives. What were the evolutionary pressures that led to this curious state of affairs? That would be the subject of a whole other book. What is especially interesting for our discussion is that the modern breeds exhibit distinct subsets of the ancestral foraging sequence and that different breeds express partial predatory sequences in different ways.

A great example of this was evident when we kept large pens of livestock-guarding dogs (Anatolian shepherds, in this case) and herding Border collies at our Hampshire College lab. We used to feed both groups of dogs with stillborn calves that we obtained from local dairy farms. The livestock-guarding dogs never exhibited the DISSECT motor pattern, which normally occurs toward the end of the canid predatory sequence. Large chunks of meat can't be swallowed whole, and successful foraging requires the prey to be torn apart before it can be consumed. When we provided calves to the Anatolian shepherds, we had to cut them open ourselves so that the dogs could feed, otherwise the calves would rot and the guarding dogs would go hungry. They simply couldn't get them open. Border collies, however, would tear open and eat dead calves with abandon.

The absence of the DISSECT motor pattern showed up in a couple of other interesting ways when we placed livestock-guarding dogs with cooperating farmers in order to assess their working behavior. One sheep rancher called us to proudly say how well his dog was working out: he had an old ewe that got sick and went off to die, and the dog stayed with the dead sheep for three days, still guarding it even after death. Shades of Greyfriars Bobby! His dog seemed to him to be the epitome of a faithful guardian. But our reaction was, "Yes, that's a good dog—because it doesn't have DISSECT." Another farmer once phoned to complain that we had given him an awful dog. One of his lambs had gotten tangled on barbed wire that tore the animal wide open. What happened? Horrors! His dog ate alive the (already "dissected") lamb. To the farmer, eating a lamb seemed like the antithesis of good guarding behavior. In fact, neither dog had any predatory motor patterns: no EYE-STALK, CHASE, GRAB-BITE, KILL-BITE, or DISSECT. Their only foraging motor pattern was CONSUME. From an adaptive working point of view, this was a perfect livestock-guarding dog—no faults.

So in some dog breeds, a part of the foraging motor pattern is simply missing—hence the overall behavior differs in quality from the wild type. The frequency of the motor patterns that are expressed can differ as well. Many of our livestock-guarding dogs would never once CHASE a ball that was thrown past them, no matter how much we tried to entice them. Border collies, by contrast, will run after anything that moves. In addition, the sequence in which motor-pattern components appear can differ between the breeds as well. The sheep dogs provide another excellent example. In a study at our Hampshire College lab, we found that in Border collies the EYE > STALK sequence is followed by the ORIENT motor pattern 85 percent of the time: if they detect and turn toward an object of interest, they are overwhelmingly likely to fixate on it and go into the stalking pattern. Livestock-guarding dogs, conversely, rarely if ever express EYE > STALK following ORIENT. Instead, this prior motor pattern releases investigative and social behaviors.

Finally, the "prey" object to which a behavior is directed can differ, too. Border collies displayed EYE > STALK > CHASE to conspecifics (other dogs) and also to different species, such as sheep. Most European sheep-guarding dogs (e.g., the Great Pyrenees, Maremma, and Anatolian shepherd) do not display these predatory sequences to non-conspecifics, or they display them only rarely. You can also see these differences reflected in the standards that dog fanciers adopt in shows and competitions. When Border collies are judged in competitive field trials for their ability to herd sheep, they are being measured in terms of whether they have "correct" breed-typical motor patterns. If a Border collie shows GRAB-BITE or KILL-BITE toward sheep, it is deemed to have a fault and is disqualified from herding competitions and tests.

In contrast, if another kind of herding dog—say, a heeler like a corgi or a Queensland blue heeler that is meant to nip at a cow's legs in order to move it—failed to show GRAB-BITE, it would be disqualified. Similarly, if a pointer goes into CHASE after a bird it just flushed, it's a fault, whereas if a retriever doesn't go after a wounded bird, it is disqualified in hunting trials. The standard for each of these breeds reflects one or more missing motor patterns compared to the full sequence that is characteristic of related wild canids like the wolves and coyotes.

Over the years, we've worked with our students to construct comprehensive ethograms that characterize the intrinsic rule systems of different breeds of dogs and to collect careful observational data on the quality, frequency, and sequence of their foraging motor patterns. We looked especially at working animals: livestock-guarding dogs like the Maremma and herding dogs, including headers like Border collies and heelers like corgis, as well as the hounds, pointers, and retrievers that are associated with modern recreational hunting. Hounds like the beagle are meant to accompany humans on a hunt and to directly participate in the chase for prey, while pointers aid hunters by detecting and keeping a fixed bead on the target, and retrievers wait until a prey animal is brought down by the hunter and are meant to bring it back. When we recorded and analyzed these data, it became clear that there are indeed significant breed differences in foraging behavior patterns (and that these differences are appropriate to their particular working tasks). Table 2 informally summarizes these results.

You can see that the livestock-guarding dogs, as we've already indicated, exhibited foraging motor patterns at very low frequencies (in some individuals, not at all) and that no one pattern necessarily triggered another. They oriented toward potential prey infrequently, and if they did, it rarely led to a succession of other predatory motor patterns. By contrast, herding dogs were predisposed to respond to prey, and once this occurred, EYE, STALK, and CHASE were inevitable results.

When it comes to training these dogs, a working-dog professional will be able to tell you very quickly whether a particular dog is trainable, and at what level. They look at which motor patterns are displayed as well as the threshold points where the animal goes from one motor pattern to the next. If a dog is missing an essential motor pattern, pro-

TABLE 2. BREED-TYPICAL MOTOR PATTERNS (FORAGING BEHAVIOR)

Breed Type	Motor-Pattern Sequence					
Wild type	ORIENT>>	EYE>>	STALK>>	CHASE>>	GRAB-BITE>>	KILL-BITE
Livestock-guarding dog	(orient)	(eye)	(stalk)	(chase)	(grab-bite)	(kill-bite)
Herding dog (header)	ORIENT>>	EYE>>	STALK>>	CHASE	(grab-bite)	(kill-bite)
Herding dog (heeler)	ORIENT>>	eye	stalk	CHASE>>	GRAB-BITE	(kill-bite)
Hound	ORIENT>>			CHASE>>	GRAB-BITE>>	**kill-bite**
Pointer	ORIENT>>	EYE	(stalk)	(chase)	GRAB-BITE	kill-bite
Retriever	ORIENT>>	eye	stalk	chase	GRAB-BITE	(kill-bite)

Note. Motor patterns that are capitalized occurred in our data at a high frequency; lowercase indicates that the behavior was relatively infrequent and might have been absent entirely. Behaviors in parentheses are considered to be faults in show and competition animals or are dysfunctional for a breed type's particular working task. When motor patterns always occurred linked together, in a particular order, their relationship is indicated by the symbol >> and the associated patterns are marked in bold.

fessionals know that there is no sense in continuing its training. And sometimes an animal will display motor patterns it should not have for its working type. Any predatory motor patterns displayed by livestock-guarding dogs should disqualify them for work. We did sometimes see individuals that displayed CHASE; once we encountered an Anatolian shepherd holding a GRAB-BITE on a sheep's hind leg while it continued to graze. Bad behaviors in guarding dogs.

If the handler notes the presence of a disqualifying motor pattern early after its onset and immediately prevents its performance, it is sometimes possible that the offending motor pattern will drop out of its repertoire. The neonatal foraging behaviors, for example, have an immediate onset at birth, but suckling behavior will offset quickly if the pup is not allowed to suckle at all. The same is true of adult foraging motor patterns: livestock-guarding dogs that are not allowed to CHASE or display GRAB-BITE will be fine in a few days. (For breeders this does bring up the question of whether such dogs should be bred. Even though the dog can successfully perform, it will still pass on the disqualifying motor patterns in its genes.)

Trial Border collies are particularly interesting in regard to the problem of exhibiting disqualifying motor patterns, early motor-pattern offset, and missing motor patterns. Trial handlers will quite often tell you that they appreciate a little "kill" in the dog because its performance will be more intense. That means they like dogs that display some degree of GRAB-BITE and KILL-BITE. These dogs will not be disqualified in trials, however, if the judges never see the display. A top handler in the ring never lets the dog cross the threshold between CHASE and GRAB-BITE, even if the dog has these motor patterns in their repertoire. That doesn't always pan out. In figure 18, a less-than-competent handler has lost control of the dog, and it has gone on to display the KILL-BITE motor pattern.

Note how perfect the performance is—it looks like an image from "wild nature" of a wolf taking down a deer. But the dog in this picture had never before tried to kill a sheep. Do you think it learned to do this so precisely? It's hard not to conclude that we are seeing a hardwired behavior, a motor pattern that this dog shares with its wild relatives.

These stories are compelling and informative, but we want to be able to quantify more precisely what we really mean by breed differences and differences in their behavior. In another of our studies at Hampshire College, conducted by our student Ellen Torop, we recorded the quality and frequency of seven motor patterns in each of twelve dogs of different breeds. One of them was the Castro Laboreiro, a Portuguese breed that is described as a livestock-guarding dog. We ran a statistical cluster analysis on them. Each dog provides a little seven-dimensional cloud of data that reveals how closely the quality and frequency of its motor patterns resemble those of other dogs. The analytic technique is used to group (or cluster) them into statistically defined types, quite independently of any prior beliefs we might have had about breed similarities. These results are summarized in figure 19.

This picture generally dovetails with our other studies, showing that there are consistent motor-pattern differences between guarding and herding dogs. That's important in its own right: there are indeed systematic behavioral differences between breeds. But in addition, this kind of careful quantification and statistical analysis can sometimes reveal unexpected results. The Coppingers had acquired the Castros

FIG. 18 BORDER COLLIES AREN'T SUPPOSED TO HAVE THE KILL-BITE MOTOR PATTERN.
AS THE SCOTS SAY THIS DOG WILL "LOSE A WEE POINT" AND SHOULD BE DISQUALIFIED
FOR PERFORMING THE MOTOR PATTERN. THE SHEEP WASN'T HURT. PHOTO BY LORNA
COPPINGER.

in Portugal, where local farmers sell them to tourists as a distinctive
local breed—and say they use them as livestock-guarding dogs (fig.
20). But you can see in figure 19 that Castro Laboreiros cluster statistically with the herding dogs. Their motor patterns are more similar
to Border collies than to Maremmas or Anatolian shepherds. And as

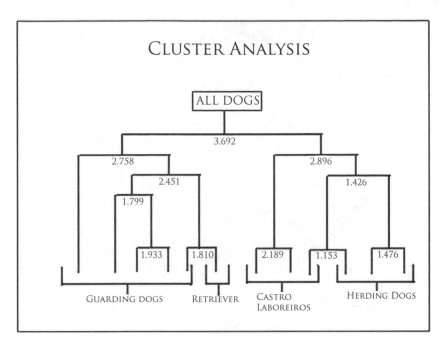

CLUSTER ANALYSIS

ALL DOGS

3.692

2.758   2.896

2.451   1.426

1.799

1.933   1.810   2.189   1.153   1.476

GUARDING DOGS   RETRIEVER   CASTRO LABOREIROS   HERDING DOGS

FIG. 19   SIMILARITY OF TWELVE DOGS OF DIFFERENT BREEDS BASED ON THEIR MOTOR PATTERNS.

it eventually turned out, when the Castros came back to our lab, we could not get them to reliably and consistently play the working role of a livestock-guarding dog. Of one hundred Castros that we put out on our cooperator program (sheep ranchers helping us to see whether the dogs would effectively control predators) only 20 percent turned out to be successful guardians—the rest were sheep chasers. This contrasts dramatically with a 70 percent success rate in other guarding breeds. In spite of what local people may call these dogs (or wish they were like), an ethological analysis paying careful attention to their motor patterns could have predicted that Castros would behave more like the herding than the guarding breeds.

The question remains whether these breed differences are truly due to intrinsic genetic properties shaped (in the case of dogs) by artificial selection. Could they instead be results of nurture—artifacts of training or consequences of environmental differences? Or are they the result of a synergism between all these forces? Professional dog breeders certainly wouldn't be surprised to find that these behavioral patterns

FIG. 20 HERE YOUNG TIM COPPINGER BUYS A CASTRO LABOREIRO PUP IN PORTUGAL AND GETS SOME INSTRUCTION ON HOW TO REAR IT. PHOTO LORNA COPPINGER.

are innate: after all the business of breeding is to intentionally select for particular traits by manipulating reproduction. Animal trainers, by contrast, are in the business of trying to induce behaviors for particular working or other purposes. Could the differences be due to training? We don't think so. The fact is, it's not possible for dog trainers to teach Border collies to eye sheep, or setters to point birds, or retrievers to retrieve. We carried out numerous "cross-fostering" experiments with Border collies and guarding dogs, including Maremmas, raised in the same early environment with each other's mother. This treatment never changed the picture of their breed-specific foraging motor patterns. Maremmas cannot be taught to work as herding dogs: intrinsic parts of the machinery that are needed to do the job are inactive or missing. And the timing of development also plays a critical role. Until a young Border collie begins to exhibit EYE > STALK it simply can't be trained to respond to commands that will allow it to manipulate the movement of sheep. These behaviors emerge spontaneously during ontogeny (i.e., the course of the life of an individual), and whether a dog of a particular breed can be trained depends on where it is in its life history.

Do individual differences, environmental factors, and developmental events affect performance? Yes—there is no question about it. A Border collie that happens to exhibit a low frequency of EYE > STALK when you try to train it is not likely to turn into a good herding dog. Training (refinement and reinforcement) can only commence after the onset of the behavior, and it won't succeed unless the appropriate "genetic" motor patterns are reliably and robustly represented in the individual's behavioral repertoire.

However, if you don't raise that set of genes in the proper environment, you won't get a good working dog either. Our experimental work has shown that there is a specific environment in which a livestock dog needs to be raised. If you don't raise the dog in that setting, you ruin its future as a livestock guardian. Not only do you ruin it for the moment, but there is also no going back and correcting the mistake. One of the greatest difficulties we have with dog breeders is that they believe their dogs' behavior is entirely hardwired and therefore inevitable—all you have to do is buy a livestock-guarding dog and it will guard your sheep from predators. We ethologists, who otherwise agree that genetic hardwiring is a crucial dimension of behavior, find ourselves frustratingly saying, over and over, that farmers also have to pay attention to the developmental context: if you don't raise the dog in the proper environment, you ruin its adult working performance. It's the nature/nurture conundrum all over again.

The evolutionary bottom line, however, whether we are looking at foraging or any other type of behavior, is that dogs and every other animal are adapted to behave in some particular environment. Their genes—that is to say, the shapes and motor patterns that the genes build—set important limits on what they can do in the world. But the world in which they behave and grow matters a great deal, too. How these forces actually interact continues to be a great puzzle and a source of fascinating argument. In the following chapters we explore in much finer-grained detail three ways of looking at this long-standing debate about the nature of behavior.

6 INTRINSIC BEHAVIOR

What could be more beautiful and natural than a newborn infant suckling at its mother's breast? You can see it as a lyrical expression of a mystical life force if you like. For the ethologist, it is a motor pattern that expresses a set of rules: Find the teat. Attach to it. Suck to create a pressure differential while engaging a rhythmic paw-pressing movement that stimulates the milk flow. This basic behavioral shape is present in many newborn mammals. The growth and activity of newborns crucially depends on the energy supplied by mother's milk, and they need it immediately if they are to survive. The behavior has to work exactly right from the very start.

These nursing motor patterns are intrinsic. Like predatory behaviors and a myriad of other behavioral characteristics they are inherent in the way the animal's body and brain are built. They are a product of its genes. The shape of the newborn's mouth and tongue is essential to the task of attaching firmly to the nipple and creating the pressure differentials that are required. If you stick your finger in a pup's mouth—one of the little sensual pleasures of life—you can feel that muscular mobile little tongue wrapping itself tightly around and holding the finger, rhythmically pulling on it as it develops a partial vacuum in its mouth (fig. 21). (Later in life, when the pup metamorphoses into an adult, the oral cavity will change in shape: it will enlarge and the tongue will flatten and become more flaccid—a shape that is useless for suckling and makes it much less fun to insert your finger into an adult dog's mouth.) The newborn's brain shape is crucial as well. The "rules of suckling"

FIG. 21 IT IS TO THIS LAMB'S SELECTIVE ADVANTAGE TO SUCKLE HOWEVER AND
WHEREVER IT CAN—EVEN FROM AN ANATOLIAN SHEPHERD DOG. BUT IT IS TO MOM'S
SELECTIVE ADVANTAGE ONLY TO SUCKLE HER OWN OFFSPRING. MANY WILD SPECIES
AND BREEDS SUCH AS OUR SLED DOGS AND BORDER COLLIES TENDED TO KILL YOUNG
PUPS THAT WERE NOT THEIR OWN. OTHER BREEDS, LIKE THE LIVESTOCK-GUARDING
DOGS, DON'T SEEM TO CARE. PHOTO BY JAY LORENZ.

are (somehow) wired into the neural pathways and the electrical and
chemical signaling mechanisms that regulate movement. There is no
question about this: the intrinsic suckling motor pattern is not learned.
It appears mere moments after birth, and it is inconceivable that any
newborn could ever have the opportunity to acquire the behavior by observation or through practice.

Moreover, many mammals, including dog pups, must suckle almost
immediately after birth. There is just a small window of time when the
animal has to attach to the teat and draw milk down. Lambs, for example, have to be on their feet and suckling within fifteen minutes of
birth. If not, the mother takes off, and it is almost impossible to con-
vince her to come back. If the lamb is to survive, it will have to be bottle-
fed—a charming thing, but in fact a consequence of motor-pattern
failure.

The situation is a bit different in dogs with large litters. If the first pup is slow getting to a teat, there are other pups coming along that do make the attachments, and the mother will stay. Nevertheless the timing is still critical for dog pups: if they don't begin to engage it early enough, the SUCKLE motor pattern drops out of the repertoire, and the infant can't be taught to do it thereafter. One of our bird dog mothers got a pup stuck between her back and the edge of her nesting box, where it spent some part of the night. We made the situation right in the morning. But to no avail—we had to milk the dog and feed the pup with an eyedropper because it was no longer capable of suckling.

Nest building is another critical part of the intrinsic birthing process in dogs and other canids. The mother marks out and manipulates the environment in some way to establish a quiet dark space away from other dogs. One of our livestock-guarding dogs, a Šarplaninac we named Sirna (the "dark one," in Serbo-Croatian), built herself a three room "apartment" in a sand bank. Her nest was a spacious cave—but it also looked like it might collapse at any moment and everybody, mom and pups alike, would be buried alive. (See plate 4 for a photo of a typical nest.) Breeders often see potential danger in allowing a mom to build a nest like Sirna's and they may try to suppress the display of nesting motor patterns. But nest building is part of the normal sequence of intrinsic birthing events in dogs: unless it is expressed, mothers can get into trouble.

At some service-dog organizations veterinarians insist that mothers should have their pups in a clean, antiseptic environment. Well-meaning, they don't want to let the mother build a nest from old dirty rags. But as a result, a mother may give birth to her puppies on a hard (if very clean) floor under bright lights, surrounded by other noisy dogs, and attended by "puppy watch" staff on hand to help if anything goes awry. It often does. Pups born in sterile environments will, for example, occasionally eat their pups rather than just trimming their umbilical cords. What has gone wrong? As we observed in our discussion of predation, one motor pattern provides the motivation for the next one in a sequence. When a dog can't express her intrinsic birthing motor patterns in the proper sequence—including nest building—the whole process can be short-circuited. Our best advice to owners of dogs about to give birth is to let the dog make or find its own nest (dirty and messy or not). Then go to bed, get up in the morning, and count the puppies.

Intrinsic behaviors are not—and cannot—be taught or reshaped by human trainers or companions. They don't require practice. They cannot be circumvented except at the animal's peril. They are neither learned through experience nor the result of factors in the animal's external environment. Intrinsic behaviors are historical products of species evolution, inherited consequences of the action of an animal's genome, and the trajectories of growth and development that the genes regulate. When all is said and done, we could sum up the story of intrinsic motor patterns simply by saying that they constitute the built-in "ticking mechanism" of a dog.

THE LOST CALL

In their 2001 book *Dogs: A Startling New Understanding of Canine Origin, Behavior, and Evolution*, the Coppingers tell the story of Lina, a Maremma livestock-guarding dog that was bought in Italy as a pup and raised at Hampshire College's Farm Center. It's worth repeating here because it provides such a compelling example of what we mean by intrinsic behavior.

Big, white, and stolid in temperament, Lina was part of our twenty-year project to study dog behavior and development (and as we've described earlier, to encourage the use of dogs for the nonlethal control of predators). She was a working animal who spent her days on grass pastures in the company of a flock of sheep who were occasionally preyed on by coyotes or beset by neighborhood dogs. Overnight she slept on a straw-covered bed in the sheep barn amid her charges. During her first pregnancy, Lina kept to her usual work pattern. She would sit on guard in the fields during the day and then return to the barn when the sheep were brought in. Lina passed the sixty-three days of a dog's gestation period with no difficulty, and she went into labor one summer day while she was still working the field.

When her labor contractions had started in earnest, Lina gave birth to the first pup, expelling the placenta and the amniotic sac surrounding the newborn. Normally—these are all intrinsic, unlearned motor patterns in a dog mother—she should have then torn open the sac and licked the pup clean, rubbing its skin with her tongue to stimulate circulation and respiration. Then she should have cut the umbilical cord with her teeth. On this particular occasion—we don't know why—Lina dropped the first pup in the tall grass, still in its afterbirth, and left it there, retreating to the nest she had made in the barn without a glance back at her firstborn. Luckily, the placenta had broken and the pup's head was out in the air.

After many hours of labor, Lina eventually gave birth to seven more pups. We might never have known about that first little pup, lying uncleaned and untended in the field except for the fact that when Lina abandoned it the pup began to emit a loud and persistent vocal signal. We call this early motor pattern the LOST CALL—although "being lost," separated from one's mother or mates, may not be the best way to understand why it occurs, as we'll soon see. The call has a very distinctive sound, rising and falling in pitch for a few seconds like a brief siren wail. All pups—and that includes the newborn of other canids such as wolves, coyotes, and jackals—give it when they're isolated as small puppies. Typically, when a pup produces the LOST CALL, the mother responds almost instantly: she detects the signal, locates its source, and rushes to retrieve the errant offspring and bring it to safety.

That's how we expect a communication system to work. A sender transmits information over some physical channel to a receiver. From an evolutionary standpoint, senders engage in this kind of motor activity because it has a fitness benefit in terms of feeding, avoiding hazards, or reproduction. The same is true of receivers: they stand to derive a benefit by detecting and acting appropriately in response to the signal's information. Like many animal signals, a behavior like the LOST CALL is clearly adaptive—it serves the critical function of allowing the newborn to survive a severe potential hazard (and the mother to maximize her reproductive success), every member of the species can do it (at the right time of life), and the behavior occurs in many animals that share a common biological history or phylogeny.

The LOST CALL has all the earmarks of a motor pattern—albeit one where the animal's movement activity isn't overt or visible or easily describable in detail by a field ethologist. Nevertheless, like any motor pattern this vocal signal is caused by a set of movements, largely internal, that can be characterized in terms of quality, frequency, and sequence.

To begin with, what is its quality? When a pup gives the LOST CALL it raises its head, forces air from the lungs with its respiratory muscles, and vibrates a set of membranes (the vocal folds) in its larynx by means of another set of fine muscle movements. This is in fact the general mechanism by which acoustic signals are produced in mammals. The energy imparted by this muscle activity is transferred to a vibrating body of air that is further affected as it moves through the shape of the animal's head. The result is a sound wave with a particular acoustic character that we (or another dog) can detect when it reaches the ear.

As we've said, ethologists pay attention to movements of the animal's body that we can detect, observe, and measure, and in the case of vocal behavior we usually can't see much of that movement directly. Happily it is possible through modern sound-analysis technology—sound spectrography—to precisely visualize and measure the characteristics of the wave that results from motor activities of the vocal tract and so to capture the quality of the motor pattern. In figure 22, you see a sound spectrogram (commonly called a sonogram) of an original recording that we made of Lina's first pup when we discovered it in the field. It illustrates the pattern of acoustic energy produced by the animal's internal movements.

Time is represented on the horizontal axis: you can see that this call lasts for a few seconds. (In fact it was emitted repeatedly for quite some time.) The vertical axis depicts acoustic frequency: the rate of vibration of structures in the vocal tract, which produces its apparent pitch. If this had been a simple "pure tone," like the sound of a tuning fork, the spectrogram would have shown only a single unchanging frequency—a single bright band of energy. The call of Lina's pup, like many biological signals (including human speech), is a complex wave that consists of simultaneous multiple frequencies. Eight of these components of its "spectrum" are visible in figure 22. The LOST CALL's wave contains multiple evenly spaced "harmonic" frequencies, a pattern that gives the sound a "tonal" auditory character, with something of a musical

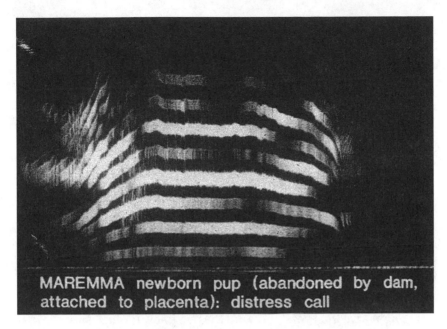

MAREMMA newborn pup (abandoned by dam, attached to placenta): distress call

FIG. 22 A SONOGRAM OF THE LOST CALL. THIS IS A COMPLEX, LOUD SOUND THAT IS ENERGETICALLY RATHER EXPENSIVE TO MAKE. A PUPPY CAN GIVE THIS CALL FROM THE MOMENT IT IS BORN, BUT BY TWO MONTHS IT IS GONE FROM THE REPERTOIRE OF VOCAL SIGNALS.

quality. (Why? See chap. 8.) The overall pitch of the signal also changes with time. As you can surmise by looking at the vertical axis, the call's pitch rises sharply at the onset, where the higher pitch is steadily maintained for some time, and then declines toward the end. This kind of relatively abrupt rising-falling pitch contour, in fact, tends to make it easy to locate the source of the sound—clearly an adaptive advantage for a "lost" call.

Finally, you can tell from the spectrogram how loud this call is—a function of the amplitude of the wave, how much energy was put into it by vocal tract movement—by looking at the relative brightness of the frequency components. Our abandoned pup clearly expended a lot of energy! So it wasn't difficult for us to find the plaintive newborn: its very loud cry was easy for us to detect and the acoustic shape of the signal helped us to hone in on the calling pup even though it was hidden in tall grass.

Lina could certainly have detected the signal as well. A dog's auditory system is very much like our own. In fact, in certain respects dogs have better hearing than humans—they are able, for example, to perceive sounds at a much higher pitch than we can. Remarkably, however, Lina seemed to pay absolutely no attention to this call. She simply went about the business of delivering the rest of her litter, and we had to go out in the field to rescue the pup ourselves. First, we made a recording of the sound. Then, we cut the cord and brought the abandoned pup back to Lina's nest. She placidly accepted and cleaned it and went on to nurse the pup she had previously ignored along with its seven littermates. Why didn't we just leave the puppy there and see if she would eventually come and get it? Because we knew what would happen: the energetically expensive call would exhaust the puppy, it could attract predators, and the puppy's suckling motor pattern would be lost.

This story raises a lot of questions about both mother and pup. First, how did a newborn just out of the womb know how to make a signal that might save it? Second, how did the pup know it was lost in the first place? Very young canid pups are highly limited in their sensory abilities. At birth, their eyes and ears aren't "open"—these systems won't be fully formed and fully functional until several months into life. Lina's firstborn, still in the amniotic sac, had virtually no experience of the external world. It had never heard a LOST CALL before. Yet the pup hit the ground and almost immediately engaged in a motor pattern that told the world, in effect, that it was out of contact with mother and in distress.

Moreover, it began to produce this behavior at a very high rate—over and over and over again, and with no immediate external reward for giving the call. Now it's no trivial matter for a tiny animal, weighing barely a half a pound, to call so loudly, insistently, and repeatedly. It takes a great deal of energy to vocalize. If you've ever just talked on and on for hours on end and then found yourself exhausted and craving a meal, you'll appreciate how many calories are burned up by the motoric demands of vocal fold vibration and its associated muscle activity and respiratory movement. For a little animal to make the huge energy investment that the LOST CALL motor pattern requires, it must be really important. And, of course, it is. For a lost or otherwise distressed newborn pup, giving the right signal at the right time is a life-or-death issue.

Luckily for the pup it doesn't have any choice in the matter. Whenever a newborn pup finds itself away from the warm and cozy nest, alongside mother and littermates, it produces the call. If you gently pick up the pup and place it on a cold metal lab table right next to the nest, it will perform exactly the same motor pattern. Put a light bulb next to the animal, however, and the vocalizing will stop. Is the pup comforted by the presence of a glowing light? Well, its immature visual system in fact prevents it from being able to see the light itself. But a newborn pup is quite capable of sensing temperature differences, and a light bulb emits heat along with light. Careful experimentation showed that when a pup is warmer on one side than another—whether the heat is generated by a mother's body, its littermates, or a light bulb—it won't give the call. When the pup is alone without a light bulb glowing to one side, where the temperature gradient is absent, the call is released. This is thermotaxis—essentially an automatic reaction in response to heat differences. (When our bird dog pup got stuck between its mother and the edge of the nest it didn't give the LOST CALL even though it was in danger, unable to initiate normal suckling behavior. Why was that? Wedged against the mother's body, it simply was warm enough.)

You can see the same sort of thermotaxic reaction even in very simple animals. A nematode worm, for example, will move along a heat gradient keeping to the warmer side. Honeybee movement in a dark hive likewise seems to be governed, in part, by thermotaxis. Like the LOST CALL, this intrinsic motor activity is released by a particular kind of sensory stimulus, a bit of information for which the animal's brain has an intrinsic response. Like any motor pattern, the expression of the behavior requires no prior experience, it is generally characteristic of the whole species, and it is stereotypical in that species. Everyone has the pattern, and when it's triggered everyone does it essentially the same way.

If you raise pups and you're diligent about keeping them safe and warm, you might watch over many generations of puppies and never observe the LOST CALL even once. Let something go wrong early in life five generations later, however, changing the temperature of an individual's otherwise comfortable existence, and the motor pattern will suddenly appear as if from nowhere. And it will be perfect the very first time. Evolution has provided every little lost or distressed pup with the necessary inherited biological machinery: the pup is wired to detect a gradient temperature change that triggers a set of intricate adjustments to its pattern of breathing and the tension of its vocal cords. Whether or not the mechanism is ever engaged—like a low gear on a car's automatic transmission that a driver may never use if conditions never warrant it—it is always available because it is an intrinsic property of the machine.

Remember the ethological dictum that motor patterns are adaptations. They are products of evolution through natural selection, which confer a fitness benefit—that is, they enhance the animal's ability to survive and reproduce. Getting rescued by your mother when you're lost and cold certainly counts as adaptive for both mother and pup. If that is right, however, why didn't Lina respond to the LOST CALL by safely retrieving her pup? Effective communicative signaling requires a sender who transmits a signal with information—"Help!"—as well as a receiver who responds by helping. Putting energy into a signal wouldn't seem to be worth the effort if a receiver doesn't act appropriately. So we might expect dog mothers to have an adaptive intrinsic motor pattern of their own, at the ready to act in response to a signal from their offspring. This is indeed what we usually see. When the LOST CALL is detected by mother, she typically engages a RETRIEVE motor pattern, attentively orienting toward the sound, moving to locate the source of the call, gently picking up a straying pup in her mouth, and returning it to the nest. Lina didn't do this. Why not?

INTRINSIC TIMING

To begin with, we have to admit that we just don't know exactly why Lina abandoned the pup in the first place. She was a first-time mother. Could the stress and discomfort of her first delivery just have thrown her for a loop? Maybe her only impulse was to get out of the open field and back to the protection of her "nest" in the barn. She may have been in a confounding conflict between staying in the field with sheep she was guarding (like the Maremma in the Abruzzi) and leaving them to return to the nest. Maybe she just didn't notice. It's an interesting question, and we'd certainly like to find an answer. (It would be difficult, however, to try to investigate this kind of problem: we don't see this aberrant response often enough to draw conclusions from direct observation, and there aren't enough animals who behave this way that we could use as subjects in a controlled experimental study). But whatever the cause of her initial failure, once Lina was back in the safety of her nest and successfully birthing the rest of her pups, why didn't she then respond to the desperate wail of her firstborn? Why didn't she seem to notice it at all?

It's possible, of course, that the mother's retrieval behavior isn't actually an intrinsic motor pattern and instead has to be learned from hard experience. This was Lina's first litter, after all. But we don't think it was a question of learning. Other things we know about dogs suggest that she simply couldn't do anything about it. For dogs, the retrieval motor pattern simply won't appear until the last puppy is born. As we've said, intrinsic motor patterns are normally released by some internal or environmental signal. RETRIEVE is triggered by two things: the LOST CALL itself, as well as an internal physical signal—a particular hormonal state in the mother. Both inputs are required to release the retrieval behavior. The regulation of the birth process involves a complex set of biochemical changes in both the pups and the mothers, and until Lina's hormonal levels were just right—until the last fetus was expelled—her retrieval motor pattern simply couldn't engage. If we had manipulated Lina's hormone state experimentally—artificially signaling that parturition was complete—we bet that we could have induced Lina to display the retrieval response before all the pups were born.

So some motor patterns kick in only at particular times in the development and life history of an animal. They have an onset, which may be right at birth, as in the case of the LOST CALL, or, like RETRIEVE, they may only become active at a later point in life. The offset of a motor pattern can be just as marked. The offset is the point at which the animal ceases to respond with the specific motor pattern. Particular motor patterns, therefore, may occur only during a certain time period in the animal's life history. Characteristic human sexual motor patterns, for example, don't appear for many years after birth—not until we reach puberty. They are released by hormonal changes (along with sexual signals from the environment) and generally persist through much of life, though in some people they do offset at some point in old age.

An ethogram, as you'll recall, describes the quality, frequency, and sequence of motor patterns in a particular species. Ontogenetic onsets and offsets of motor patterns need to be taken into account in the ethogram as well when we characterize species-specific behavior. Species and breeds that exhibit similar (even homologous) motor patterns can have different onset and offset times. Like dogs, rat pups also produce a LOST CALL, and rat mothers have a RETRIEVE motor pattern. Rats, however, are very different, and pregnant mothers will even retrieve another female's lost pups weeks before the birth of any of her own offspring. But when Lina heard her firstborn calling, the onset time for retrieval—in dogs—had not yet been reached and she could not display the behavior.

The dog mother's RETRIEVE response to the LOST CALL is a great example of the importance of timing in onsets and offsets. Once parturition is complete, the dog mother's RETRIEVE motor pattern is active in livestock-guarding dogs for only about thirteen days (though there is some variability between breeds; a friend of ours, Kirsty Peake, did the experiment with her Yorkshire terriers and found that mothers would stop retrieving pups after only ten days). We have plenty of systematic observational data to back up the offset timing of the retrieval motor pattern, but the following anecdote tells the story well.

Ray was getting ready to go out to dinner one nasty rainy winter night when he heard the cry of a lost pup from the dog yard. He went out to investigate and found Tilly—one of his all-time great sled dogs who had given birth to a litter two weeks before—sitting placidly in a nesting box with her other offspring. Right next to the nest, only inches away, was a cold, wet, and shivering pup that had somehow fallen out into a puddle of ice cold water. Not being warmer on one side than the other, it was calling loudly and repeatedly—but Tilly wasn't paying the slightest attention. Wasn't she cognizant of her pup's obvious distress?

Why didn't she just reach out and rescue the poor little thing? Ray remembers thinking, "Tilly, what are you doing? Couldn't you just reach out and pull the pup in? What has happened to your retrieval motor pattern?" Tilly's answer might well have been, "Hey boss, thirteen days is up. I don't do puppy retrieval anymore." Put a bit less anthropomorphically, the retrieval motor pattern had reached its offset cessation point and was no longer active. Tilly's lack of response wasn't a failure of motherly love—there is no reason to think that her behavior had any psychological content at all. It was simply governed by the timing of a motor pattern—the ticking away of an intrinsic part of her mechanism.

Another incident with the LOST CALL demonstrates why it can be so tricky to make mentalistic assumptions about what might be going on when an animal is behaving. A group of our students carried out a "play-back" study using a small microcassette audio recorder with a recording of a puppy's LOST CALL. The subject was a Border collie named Flea, with a litter of seven-day-old pups in the kennel. Flea was well within the timing window in which RETRIEVE remains active. When she heard the LOST CALL being played on the tape recorder, Flea would engage the motor pattern—turn toward the sound, bound up the stairs, push a door open, and retrieve the tape recorder. Gently holding it in her mouth, she brought the little metal and plastic recorder back to her nest and placed it—still playing the LOST CALL—with her puppies.

Our anthropomorphic impulses are confounded. Didn't Flea know what she was doing? Couldn't she count her own pups and realize none were missing? And when she found the source of the signal, couldn't she tell that it wasn't a living pup at all? No. Retrieval behavior in dogs may seem to us to have the unmistakable look of maternal concern and care. But Flea clearly wasn't behaving out of "mother love." Much as we might like to think that dogs have emotions and motivations like our own, it seems clear in this case that it was the turning cogs and wheels of the machine—ultimately the genes—that made her do it. She was simply acting out an intrinsic stereotyped motor pattern.

In highlighting this picture of automatic intrinsic response we don't mean to suggest that the behavior of animals cannot be altered. Our machine metaphor notwithstanding, dogs aren't fixed automata relentlessly carrying out preprogrammed routines and nothing more . Certainly they can learn to do a great deal that isn't "written in the genes"; to tell the full story of animal learning would need another book or two (or three). However—and this is a very important qualification—what any animal *can* learn to do is limited by its shape and the intrinsic motor patterns that underlie its behavior.

INTRINSIC THRESHOLDS

People who train working dogs understand these limits. Training a dog depends critically on the presence of particular motor patterns. Until a Border collie expresses the EYE > STALK > CHASE sequence of motor patterns, there is nothing a trainer can do to make the animal learn to be an effective herder. Until a pointer expresses the intrinsic POINT motor pattern, it is effectively untrainable for its job. Border collie performance trials provide a clear example of this. As the dog approaches the sheep, it displays the EYE > STALK pattern. At some point, it has to leave EYE > STALK and go into the CHASE motor pattern. (By the way, CHASE in Border collies has a shape all its own—it is circular and not straight on, as in other dogs and wild canids.) The point where the dog leaves EYE > STALK and goes into CHASE is called the balance point by handlers. We call it a threshold (fig. 23). Once the dog passes the threshold, the next motor pattern in the sequence is released.

That threshold point, in fact, varies a bit between individual dogs. Handlers who are primarily interested in trials in closed rings, before an audience, prefer animals with very short thresholds. Those of us who want to herd sheep on the open range prefer dogs that have long thresholds that you can see a mile away. But no one, to our knowledge, has ever been able to change the threshold distance on an individual dog: if you don't like the your dog's balance point—get another dog.

To put it a bit paradoxically, a trainer has to breed animals for the right theshold. If you're thinking of competing in competitive show trials, buy a dog or a pup from trial stock; if you intend to herd sheep in the mountains, buy a dog or its puppies from a shepherd. The threshold point is a hardwired property of the dog's shape—it is a fixed genetic trait. When it comes to winning a herding trial, it's a waste of

FIG. 23 MOTOR PATTERNS ARE PERFORMED IN SPECIFIC ENVIRONMENTS IN RESPONSE
TO SPECIFIC SIGNALS. A GOOD BORDER COLLIE HANDLER KNOWS WHERE THESE
THRESHOLDS BETWEEN THE MOTOR PATTERNS ("BALANCE POINTS") ARE LOCATED.
IF YOU WANT THE DOG TO GO FROM EYE STALK TO CHASE, YOU LET IT CROSS THE
ENVIRONMENTAL THRESHOLD. PHOTO BY LORNA COPPINGER.

time to think that you'll be able to make a Border collie do your bidding
simply on the strength of your personality, the cleverness of your training technique, or your dog's own ingenuity. The breed may be high on
the list of "intelligent" dogs in the public imagination. But in the end,
a Border collie's behavioral capacities are fundamentally a result of its
repertoire of intrinsic motor patterns and its genetically determined
shape. That's what we think is true of all dogs—and of all animals.

7 ACCOMMODATION AND BEHAVIOR

The picture of intrinsic behavior in the previous chapter is as close as we're likely to come to Descartes's conceit of an animal as a ticking automaton. Indeed, we do think that much of the behavioral character of a dog is in large measure the result of its intrinsic properties—how the animal's mechanism is designed (by evolution) and built (by its genes). No matter how development proceeds, no matter how much learning takes place, no matter what the animal's inner mental life might be, dogs inherit physical and behavioral shapes that never vary significantly between individuals or change between generations.

But the intrinsic shape of the animal isn't by any means the whole story of behavior. Dogs and other animals are not simply mass-produced mechanical devices that leave the factory with a manufactured shape and then never alter their form. They grow—and many of their growing shapes and behaviors change, sometimes quite dramatically, over an animal's lifetime. It is true that in the course of growth some parts and properties do remain fixed and constant. But others are in fact modifiable: we call these accommodations.

The term "accommodation" has been used in many ways by developmental biologists and psychologists, but we first came across it in the work of early twentieth-century embryologists. In order to illuminate our conception of behavioral accommodation, therefore, we'd like to first take a moment to think about some fundamental insights from embryology, the study of early growth and form.

Every multicellular organism starts life as a fertilized egg. That initial cell contains all of the organism's genetic information. It then divides and redivides to become a larger collection of cells. Very early on, the blastula arises, a hollow sphere of undifferentiated cells. This sphere of cells soon does begin to differentiate and develops a primitive streak—some of the blastula cells line up to form a structure that will determine the differentiated shape of the embryo as it grows. The cells of that primitive streak guide the development of the gastrula, with its three enfolded germ layers: the ectoderm, mesoderm, and endoderm. Cells in each of these layers result in specific tissues—and ultimately, organs—in the developing embryo. The ectoderm produces skin cells (epidermis) and neural crest cells that will later develop into the brain and nervous system. The mesoderm gives rise to somites that form muscle tissue, blood and blood vessels, bone, and connective tissue. The endoderm produces the lining of internal organs, the digestive and respiratory systems, and other organs associated with digestion, including the liver and pancreas. At the gastrula stage, many cells have become fated—they "know" what they are going to be when they grow up. When the animal is still a little round three-layered ball, the cells that are going to become an eye are destined to play that role. They don't have a choice in the matter.

Just how all this differentiation and specialization comes about is one of the great mysteries that embryologists and geneticists are working hard to unravel. It probably has to do with the location of cells in the embryo and the activity of neighboring cells. But it is indisputable that certain cells come to have inherent properties that will result in specific outcomes in a growing body. It is these properties that the early twentieth-century embryologist Victor Twitty called intrinsic—and his usage, in fact, is the origin of the term we prefer to use in place of "innate" or "instinctive."

At every point in development, however, parts of a growing body are affected by what is around them—internal factors such as adjacent cells, tissues, and organs, as well as external effects of the environment outside the body. Nutrition acquired from the outside world, for instance, is essential if cells are to continue to function, grow, and multiply. Temperature, chemical states, and a myriad of other physical properties and events in the external environment around an animal can influence the activity of cells. Moreover, the developing organism itself constitutes a complicated set of internal environmental influences. Hormonal events inside the body can have a profound effect, and as cells multiply and larger body parts develop, they may interact and "push" on one another in many ways as the animal grows.

These factors result in what Twitty called accommodative outcomes: changes in shape and structure that aren't fixed consequences of the intrinsic properties of a growing organism. Rather, they arise from the contingent action of external forces on cell division and tissue formation. Ultimately, the whole phenotype of an animal, the shape with which it can move in space at any given time—how it behaves—is a result of how its intrinsic properties accommodate to these other forces.

Twitty himself did groundbreaking research on the developmental trajectory of, among other things, vertebrate eyeballs. He observed that in a given species, every individual's eyes will always grow to the same size and shape. It's a good Darwinian bet that this shape is an adaptation to the particular visual needs of the species. Twitty concluded that the eyeball (more properly, the fated cells that give rise to it) is an intrinsic character. Eyeballs don't exist in a vacuum, of course. They need to fit into skulls: their proper function depends on the fact that they sit within and are attached by musculature to a supporting bony structure called the orbit. That orbit has to be just the right shape to do its job. So doesn't it stand to reason that its design must also be an intrinsic adaptation?

Not according to Twitty. He worked with two closely related varieties of the *Ambystoma* salamander—one large species, one small. Not surprisingly, each species normally develops an appropriately sized eyeball. In an elegant and clever experiment with embryonic salamanders, Twitty surgically transposed the cells that were fated to become an eye in a smaller animal into a larger one—and vice versa. If eyeball growth is indeed intrinsic, you'd expect the small salamander to develop a much larger eye than usual, and that's exactly what happened. If orbit size was also intrinsic—appropriately small in the smaller species—you might well predict that the experimentally induced bigger eyeball wouldn't fit into the skull. Yet it did. The growth pattern of the orbital space in the skull around the developing larger eyeball, Twitty showed, accommodated to the bigger-sized eye. The shape of the orbit itself, he concluded, isn't intrinsic—instead, it responds flexibly, by a mechanism of tissue induction, to the intrinsic shape of the eye. Moreover, the muscles that controlled the eyeball had to change in length and robustness in order to accommodate the larger eye. So did the size of the optic tectum region of the salamander's brain, which needed additional neural tissue (or a richer set of neural connections) to accommodate the larger eye. And so the whole skull itself ended up with a different shape because it had to accommodate not only a differently sized eyeball but a differently shaped brain as well.

ACCOMMODATION, SHAPE, AND BEHAVIOR

If the shape of the growing salamander changes, then its behavior will change too: the shape of an animal defines and limits how it can move in time and space. Yes, there are adaptive intrinsic (genetic) patterns of behavior—the "instincts" that were the focus of the work of classical ethologists like Lorenz. Many motor patterns are indeed stereotyped and unmodifiable, and we overlook these "fixed" aspects of shape at our peril when we try to understand the general nature of behavior. But an animal's shape is dynamic—it changes over the course of genetically guided development and it can be altered both by the animal's external environment and by the growth of parts of its own body. And as shape changes, so does behavior.

So a dog runs like a dog because it is shaped like a dog. A greyhound may be a fast runner but it isn't able to behave like a good Iditarod sled dog because the shape of its gait isn't appropriate to the mechanical and energetic demands of sled racing. The overall shape of the greyhound and the normal character of its movement are ultimately a function of its genes, and we can safely predict that the offspring of a male and female greyhound will be a poor choice for a sled-racing dog. Two Maremma parents will produce pups with a characteristic behavioral shape—which is, as we'll see shortly, missing most of the predatory motor-pattern sequence that we observe in other canids.

The general plan of a dog breed's shape (indeed of any species' over-all shape) is certainly intrinsic. But the actual shape of any individual dog, its phenotype, is in fact not strictly genetic. Animals develop by epigenesis. One definition of this word—closely related to our use of the term "accommodation"—is that the construction processes that are initiated by genes that build and run a body are themselves affected by the organism's environment and development. The dynamic inter-action between the action of genes and the environment in which they act causes new shapes to arise. New shapes mean new behavior, and a great deal of behavior results from these epigenetic accommodative processes. Indeed, growing organisms are always affected by these forces.

Consequently, no two individual animals, even members of the same species, will ever have exactly the same shape. Consider the fact that identical twins, sharing an essentially identical genome, don't look exactly alike. Why not? Because they can't grow up in exactly the same environment. There are countless cases of otherwise "identical" human twins who are separated at birth and grow up with different families in different circumstances—and as adults they have different sizes and shapes. Here's a simple illustration. Take a pair of identical boy twins. Their father wants them to be Olympic weight lifters when they grow up, but the mother wants them to be Olympic swimmers. So the par-ents compromise, each taking one child, and from early on, they train them for their adult athletic specialties. If they're successful, one twin will look like a weight lifter. The other will have the shape of a swimmer. They'll have different proportions of muscle mass to fat and bone, and their bones themselves will be shaped differently. They'll probably have slightly different heights and may have significantly different weights. They'll utilize energy differently. And these shape differences will affect their behavior. Our hypothetical twins are likely to walk with different gaits and they'll be able to run at different speeds—resulting in a cascade of further accommodative effects that feed back on shape and behavior in countless ways.

Of course, our twins will still have many characteristics in common, since certain intrinsic properties of shape (as Twitty showed) are fixed and are rarely if ever affected by the developmental environment. So, for example, identical twins almost never vary in the size, shape, or color of their eyes. But the fact is that plenty of other organismic traits are subject to accommodation—and this will ultimately give rise to significant behavioral differences. Diet and nutrition, for instance, can have a profound accommodative effect on shape and behavior. Sled-dog racers know that you shouldn't feed growing puppies on too rich a diet—it tends to grow the long bones too fast. The bones of the dog's limbs will be long but not very thick. Most importantly, they won't have robust trabeculae, the spongy portion of bone without which fast-growing long bones are much more likely to be fragile and to break more easily. As a result they won't be able to take the grueling impacts of marathon racing; long and thin-boned animals won't behave like effective racing dogs. Good racers will instead have been fed a diet that causes the bones to grow more slowly but stronger—a shape that better enhances successful racing behavior. Needless to say, human phenotypes are just as susceptible to nutritive effects. Our twins may end up with very different shapes if their diets happen to differ or if they're fed differently in order to meet the differing metabolic demands of swimming versus weight lifting; they may well lose their chances of Olympic gold if they eat unwisely or too much.

Essentially the same thing is true for a multitude of interactions between the environment and the organism. Intrinsic growth rules certainly guide a human body toward a species-typical shape that typically includes two legs of roughly equal length. But suppose a young child whose skeleton is still growing breaks its leg badly and is immobilized in a cast for twelve weeks. When that cast comes off after three months of inactivity the child is likely to have one shorter leg and a little less brain to run that leg. The behavioral consequence will be an atypical gait, a limp. Yes, our genes provide us with an intrinsic plan for a shape that—under the right environmental conditions—causes us to walk in a species-specific manner. Change those conditions, force the body to accommodate to some particular contingency in life, and the shape of behavior will look quite different.

Twitty understood that the development of any organism is an interplay of intrinsic and accommodative growth—and, in our terms, the overall picture of an animal's behavior is likewise an interaction of intrinsic rules and accommodative alterations of those rules. As we've said several times, we like this terminology because it helps us to get away from the old and contentious dichotomy between (genetic) innateness and the effects of (environmentally induced or learned) development. That distinction, as we've said before, is naive. One might assume that the shape of a normal salamander's eye—that is, the complex structure of eyeball and bony orbit—would be an obvious candidate for simple genetic determinism. Twitty showed that this view is wrong. The salamander eye is indeed a fixed intrinsic character, determined by a "rule" in the genes of embryonic eye cells that are fated to grow to a particular size. The rules that guide the growth of the bony orbit and the shape of its skull are no less genetic—the proper genes must act to produce the right types of bone, muscle, and neural tissue. These rules don't produce a fixed intrinsic shape: they are open to accommodations to the effects of the intrinsic rule that guides the growth of the eyeball. Looked at this way, accommodative outcomes are just as genetic as fixed intrinsic factors are. Indeed, accommodation—the ability of the organism to respond to environmental influence and change—might well be regarded as a more sophisticated and complex kind of genetic mechanism than we see at work in the fixed outcomes of intrinsic growth.

Our students at Hampshire College have done some remarkable work on the interplay between intrinsic properties and accommodative interactions. Richard Schneider, now at the Department of Orthopedic Surgery at the University of California, San Francisco, carried out a series of experiments where he transplanted neural crest cells from a very early quail embryo, destined to become a quail's beak, into a duck embryo. The duck grew up with a quail's beak. In effect, Schneider replaced an intrinsic duck growth rule with a quail growth rule. The same set of genes was being implemented in a species-specific way (fig. 24). Would the quail beak fit the duck's skull? In fact it did. All of

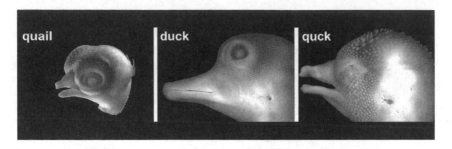

FIG. 24 TRANSPLANTING EMBRYONIC CELLS FROM A QUAIL TO A DUCK PRODUCED THIS "QUCK," A DUCK WITH A QUAIL BILL. AMAZINGLY, THE QUAIL BILL WAS ACCOMMODATED BY THE REST OF THE DUCK'S FACE: IT FITS AND WORKS. PHOTO BY RICHARD SCHNEIDER.

the necessary bones, nerves, and muscles had accommodated to the transplanted beak, and they fit together to make a functioning animal: Schneider calls it a "quck."

Now the general principle that shape determines behavior dictates that this odd "chimera" should act like a duck with a quail's bill. Of course, the shape of an experimentally modified quck isn't an adaptive product of natural selection and, like a little salamander with the experimentally induced big eyeball, the quail's new bill structure isn't actually adapted to any particular real environment. If qucks were allowed to live in the normal environment of a quail, the new shape would very likely be maladaptive. In nature, the new bill shape might interfere with normal foraging and feeding patterns or with the chimera's chances of finding a mate and reproducing. Nevertheless, it is remarkable that the genome seems capable of constructing a functional organism with a shape that responds to the exigencies of individual developmental processes—even when a particular phenotypic outcome may turn out to be less than optimal from an adaptive point of view.

Kathryn Lord investigated morphological and behavioral differences between bottle-raised and mother-nursed German shepherd pups. She found that their head shapes differed by the time they were five weeks old, just a short while before they shifted to eating solid food: the snouts of bottle-raised pups began to elongate before those of pups nursed by the mom. Lord concluded that this was an anatomical accommodation to the demands of slightly different suckling environments. Does it make a behavioral difference? Yes. Mother-reared pups showed a significantly earlier onset of the intrinsic EYE motor pattern than hand-reared pups. What will be the consequences of this change? Earlier EYE may mean that a mother-reared dog will pay more attention to its food dish or fixate more readily on other objects in its environment. It's hard to say whether such differences in adult behavior would, in fact, make any significant difference in the lives (or reproductive success) of pet dogs—human owners are disposed to cherish their pets regardless of their little quirks. The fact remains that small shape changes were correlated with behavioral differences. Under the right (or wrong) circumstances, these accommodative interactions could have a profound impact on an animal's success in life.

You don't have to experimentally induce shape changes in order to see how variety in head size and shape in dogs might affect their behavior. Over the past few hundred years, the process of creating dog breeds by artificial selection has provided many good examples of what can happen. Think of the pug, which was intentionally selected for extremely short nasal bones—a very short and very flat face (fig. 25).

The growth patterns and shape of the nasal bones are clearly intrinsic. In the course of growth and development the adjacent maxillary and premaxillary bones of its upper jaw are forced into rather dramatic accommodating shapes in order to attach those nasal bones to the palate. The resulting overall shape is bizarre even by dog standards. It works—in the sense that the animal survives and grows up with a complete face. However, the pug's accommodative facial configuration affects the shape of its nasal cavity, for instance, and breathing problems are common in the breed. Difficulty in breathing has a profound effect on an animal's metabolic capability, hence its ability to move—to behave. The facial structure also affects the shape of the oral cavity and the dog's capacity to move its tongue within the mouth, so these dogs can have difficulty in regulating their body temperature by panting. In the wild, such physiological deficits—a function of difference in shape—would impose severe constraints on the frequency and intensity with which the animal could display its normal intrinsic motor patterns. The pug has a shape that almost certainly would be inadequate if these little dogs had to forage and make a living on their own. Fortunately for pugs, we're willing to keep breeding and feeding them.

FIG. 25 A PUG FACE THAT "WORKS" (JUST BARELY). WHETHER A DOG'S FRONTAL BONE STRUCTURE—ITS FACE—IS EXTREMELY SHORT, LIKE THE PUG'S, OR ELONGATED LIKE A BORZOI'S, IT MUST STILL ACCOMMODATE WITH THE REST OF THE SKULL IN ORDER TO MAKE A "WORKING HEAD." PHOTO BY DONNA ANDERSON.

Even the way that an animal perceives the world in the course of its development can affect its shape and behavior. As we saw with Twitty's salamanders, eyeballs have an intrinsic shape and size, and the bony orbit surrounding the eye had to accommodate to experimental switches in eyeball size in the embryo. Remember that the little salamander, for instance, also had to remodel its brain tissue in order to accommodate visual signals coming from a larger eye. This is generally true of vertebrate vision: the final shape of the brain is an accommodation to what the animal sees and how much it sees when it is growing up.

David Hubel and Thorsten Wiesel won the 1981 Nobel Prize for discovering that if you raise an animal in a visually deprived environment its brain will not develop species-typical visual capacity. Nor will their eyes themselves grow normally. Take a newborn puppy and cover one eye with a patch. Leave it on for a year, then take the patch off and look at the eye—if it has an eye; it may not have developed at all. What's certain is that the pup won't be able to see normally. The animal does have an intrinsic rule to grow an eye—but in order for it to grow and the visual system to function normally, that eye, and the brain it is connected to, need to see, just as the leg needs to walk in order to grow. The patch over the eye cuts visual signals from reaching the brain, and the brain doesn't grow the neural shape in its visual cortex that is required for normal vision. The brain accommodates—albeit to an impoverished perceptual environment—and behaviors that depend on visual input will be altered as well.

Accommodation to the visual environment in the course of development can have other, perhaps even more surprising consequences. Hubel and Wiesel put very young kittens in experimental settings that limited their visual input to only horizontal or vertical lines. Normally, a kitten's brain contains specialized neurons that are responsive to these specific kinds of stimuli. After three months, the brains of the goggled kittens had accommodated (if maladaptively) to limited perceptual input. Kittens with vertical-only goggles, for example, could not reliably detect horizontal lines like the edges of tables.

A very similar developmental outcome may be seen in the behavior of some of the service dogs that assist handicapped people. While a good number of dogs do turn into effective and useful companion animals for the disabled, a relatively high percentage of them wash out. One reason for their failure, report many trainers that we've worked with, is that they balk at climbing stairs, stumble at curbs, or can't handle a grate on a sidewalk (fig. 26). We suspect that the explanation for this, at least in part, is the kennel environment in which they grow up. A kennel may be clean and spacious enough—but few of them contain the kind of surfaces and three-dimensional structures that would have provided them with an environment rich enough for their vision to develop fully (fig. 27). Not only do they lack early experience with physical structures such as stairs and curbs, it's likely that their horizontal-detector neurons simply won't be sufficiently stimulated. If there are few enough horizontal lines in the puppies' world, they may grow up effectively unable to see horizontal stair edges and street curbs, let alone negotiate them.

FIG. 26 THIS NEW GUINEA SINGING DOG WAS RAISED IN A KENNEL AS A PUP AND HAD
TROUBLE NEGOTIATING STAIRS LATER IN LIFE. HERE, THE DOG COCKS ITS HEAD TO GET
A DIFFERENT VIEW OF THE HORIZONTAL TREAD, SUGGESTING THAT IT DOES NOT SEE
HORIZONTAL LINES VERY WELL. PHOTO BY KATHRYN LORD.

Hubel and Wiesel's results suggest as well that there is a critical period for the development of vision. The notion of critical periods— similar to an idea originally proposed by Lorenz—is that there are specific times in the life history of an animal, usually early on, when it is especially responsive to particular environmental conditions. Some researchers use the term "sensitive period," but we think that misses the point. What makes a particular period in development "critical" is that

FIG. 27 MANY AGENCIES THAT RAISE PUPS UNDER STERILE (IF OTHERWISE IDEAL) CONDITIONS IN BARE KENNELS OFTEN FIND SEVERE BEHAVIORAL PROBLEMS IN THE RESULTING ADULTS. PHOTO BY KATHRYN LORD.

it is a time when an animal must grow a particular shape if it is going to be able to carry out important biological functions. If an organism doesn't grow functioning lungs by a specific time in development it can't do it later, or learn to breathe later in some other way. There is an absolutely critical period for growing a lung.

In the same way, the right kind of experience at the right time is absolutely necessary—critical—if an animal is going to exhibit species-typical behavior. Critical periods can be understood as intrinsic species-specific stages of growth—or perhaps more precisely as time-bound intrinsic constraints on development. And, like individual motor patterns, they exhibit onsets and offsets in developmental timing. Lorenz's famous imprinting studies are a classic example of this phenomenon. He showed that many young birds (such as one of his favorite study animals, the greylag goose) normally form a strong social bond with their parents, usually the mother, staying close to and following her as they feed and avoid hazards. This behavior plays a role in species recognition as well, a capability that figures importantly in later behavior. But

Lorenz found that chicks could fix on any large moving stimulus that happened to appear during a brief window of time—a critical period in the first day of life. He was able to substitute himself for the normal parental stimulus and induce chicks to follow him as if he were a parent. The drive to bond is intrinsic. Who or what the animal bonds with is a time-dependent accommodation.

Critical periods play a central role in the unfolding of behavior in many animals—including people. Human language is a great example of the interaction of intrinsic properties and accommodative effects during a developmental critical period. Many linguists believe that for children to learn their first language fluently, automatically, and rapidly, without any instruction, they must be exposed to it during a critical period that is thought to begin at around six months into life and to end at the start of sexual maturation, or puberty. Remarkably, young human children can effortlessly learn any (and often several) of eight thousand or so very different languages—as long as they are exposed to them during the critical period. When the critical period offsets at puberty, this kind of natural and effortless language learning is no longer possible for most people. You can learn a language after the critical period closes, but only with careful teaching and great intellectual effort.

Noam Chomsky, whose ideas about linguistic structure have been seminal in the contemporary study of language, brain, mind, and behavior, characterizes the basis for this general language-acquisition ability as a "mental organ." Indeed, he is inclined to refer to language growth, rather than learning or development. In this sense, he's talking just like an ethologist. The growing language organ is shared by all normal humans and it underlies our species-general ability to learn languages. The shape of this organ—that is, the neural configuration that underlies the linguistic capacity—is intrinsic. On Chomsky's view, this shared intrinsic shape gives rise to what he calls a "universal grammar"—the hypothesis that all human languages have core properties in common even though they may superficially appear to be quite diverse. That very linguistic diversity, however, demonstrates that the intrinsic human linguistic capacity must also be subject to accommodation during the critical period.

The specific quality, frequency, and sequence of linguistic input that occurs in a child's actual environment—the characteristics of language behavior to which the learner is exposed—is all-important in determining which language is acquired. A young child of Kikuyu-speaking Kenyan parents who immigrates to New York City will effortlessly acquire the English variety of that community (and perhaps Kikuyu as well) exactly like any child born into a family that had lived there for many generations. An American child who grows up in Nairobi will likely learn Kikuyu (and English and Swahili). Speaking a particular given language is, in effect, an accommodation to our cultural and social environment.

If for some reason a child gets no appropriate environmental input during the critical period, we might well expect normal language acquisition to fail entirely. One reason why input might be unavailable is that the intrinsic mechanisms for acquiring it can be damaged—as, for instance, when a child is congenitally deaf. Acquiring a spoken language, after all, requires an active auditory system that can perceive and process environmental input. A profoundly deaf child who receives no remediation (hearing aids or a cochlear implant) and no special training (lipreading, for example) will not develop speech—there is no input that can accommodatively change the shape of its intrinsic language system. (Deaf children may, of course, receive input through other channels. If their vision is intact, they can develop a manual-gestural sign language, and many deaf children certainly do.)

Accommodation to environmental input, then, seems to be a sine qua non for language to emerge. Remarkably, however, there is evidence (due to Susan Goldin-Meadow and her colleagues at the University of Chicago) that some profoundly deaf children who are not exposed to any kind of linguistic input during the critical period—neither a spoken nor a manual-gestural signed language—can still develop limited communication systems with at least some of the basic structural properties of natural languages. This unusual outcome suggests that the intrinsic shape of the human brain is driving language behavior even when the usual opportunity for accommodation to a normal linguistic environment is lacking—much like our congenitally deaf Border collies who displayed essentially normal barking.

FIG. 28 FOUR OF THE GREAT DOG BEHAVIORISTS. JOHN PAUL SCOTT (*MARKED BY ARROW*) IS IN THE CENTER, WITH BONNIE BERGEN ON HIS RIGHT AND BENSON GINSBERG ON HIS LEFT. THE LADY WELL BACK AND TO THE LEFT OF BEN IS MARY-VESTA MARSTON, WHO WROTE THE SEMINAL PAPER ON THE CRITICAL PERIOD FOR SOCIAL DEVELOPMENT WITH SCOTT.

In these respects, the acquisition of language in humans is not unlike the development of social behavior in dogs. Since the mid-twentieth century, with the publication of John Paul Scott and John L. Fuller's now-classic book on the genetic basis of dog behavior, biologists have understood that canids have a critical period for social bonding and socialization—how animals recognize and interact with one another (fig. 28).

Just how long it lasts, and what animals are sensitive to during this period of growth, remain the subject of lively debate and research. Basically, it's a matter of how dogs come to deal with other animals as well as novel objects and events in their environment. Exposure to novelty during the critical period results in long-term familiarity and a lack of (or reduced) fear or avoidance responses. Socialization in the critical period no doubt also plays a role in the development of proper species identification in wild canids. And in captive animals (and house pets), exposure during this period to other species such as humans can result in interspecies social attachment, not unlike the imprinting phenomenon that Lorenz discovered.

Kathryn Lord has argued that the critical period for socialization in dogs begins at four weeks with the ability to approach and investigate environmental novelties and ends at eight weeks with the avoidance of novelty. Scott and Fuller suggested that the offset of this critical period in dogs is later, perhaps at twelve weeks. Their conclusion may be due to differences in the breeds that were studied, idiosyncratic individual differences, or simply differences in observation and measurement.

Interestingly, Lord found that dogs and wolves alike have a four-week critical period for socialization—but the onset in wolves is two weeks earlier than in dogs. "In itself," Lord says, "the earlier progress of wolves than dogs through the critical period for socialization does not explain . . . behavioral differences between dogs and wolves." But she also realized that the two canids develop their sensory capabilities (vision, audition, and olfaction) at the same time. They can smell very early in life, though sight and hearing are not fully mature until about six weeks. This is in the middle of the dog's socialization period—but after its offset in wolves. "The consequence of this," she observes, "is that dogs began to explore the world around them at four weeks of age with the senses of sight, hearing, and smell available to them, while wolves began to explore the world at two weeks of age when they had the ability to smell but while functionally blind and deaf. . . . During the critical period for socialization, therefore, dogs have all of their senses available, while wolves must rely primarily on their sense of smell, making more things novel and frightening as adults."

These fearful responses can be mitigated only by very close, intensive, and continuous interaction with humans (presumably providing a sufficiently rich olfactory experience) during the wolf's early critical period. In the same vein, when Lord studied dog pups raised in kennels who have little or no direct contact with people for the first eight weeks of life, she found that, like wolves that haven't undergone intensive taming, they also generally tend to be very wary of human beings.

Without the right social input at the right time, these dogs become "spooky" and may also be shy of new sights, smells, and sounds as well as people.

Spookiness in dogs is a real problem when you try to train them for a specific task. Gun dogs, for instance, intended to be used as companions or assistants by human hunters, are likely to be literally gun-shy if they haven't already been exposed to the sound of gunfire or similar loud bangs during the critical period. We've found that some potential service dogs will spook uncontrollably when they hear a bus backfire. As we noted earlier, some of them are fearful of tile floors or stairs; others can't be taught to cross an iron grating in a sidewalk. It is rare that even an expert trainer will be able to induce those dogs to ascend or descend stairs with confidence. We bet that these individuals—who are likely never to turn into effective service animals—simply weren't exposed to the appropriate novel stimuli during the critical period, even though they were in all likelihood carefully and lovingly raised under the (otherwise) best of environmental conditions.

A disturbingly similar accommodative impact of the environment on human behavior is dramatically demonstrated by a decades-long study of abandoned children who were raised from birth in Romanian orphanages. These orphans were provided with adequate nutrition and meticulously clean surroundings and no doubt treated well, but staffing and funding constraints meant that individual children had only limited interactions with adult caregivers. It's not clear if humans, like dogs, have a time-bound critical period for social bonding. It appears, however, that good food and hygiene, at least, aren't enough to allow normal species-typical adult behavior to unfold. Early differences in environmental input turned out to have a profound effect on cognition and behavior in the Romanian orphans. They differed from other children on many measures related to general intelligence, they were significantly impaired in their ability to integrate sensory information with motor activity, and their behavioral interactions with other children and adults were far from social norms. Like the dogs and kittens in Hubel and Wiesel's studies, they had visual problems as well, especially in depth perception.

Comparisons with children raised in normal family settings (with relatively rich social interaction between children and adults) showed that the brains of these institutionalized children were in fact 15 percent smaller than the brains of parent-raised children, and they showed significantly reduced levels of certain kinds of cortical activity. These accommodative outcomes are highly maladaptive in modern human society, so it's fortunate they won't get passed on to future generations: accommodations aren't "genetic" in the sense of being heritable. But in another important respect, we *are* still talking about the genetic basis of behavior. Like Schneider's ducks that developed a working shape even when they needed to accommodate to genes growing a quail's bill, or the odd-shaped pug, the Romanian orphans had genomes that were capable of putting together a functional shape (even if limited and atypical of the species) when faced by unusual pressures from the external or internal environment. This capacity for flexible accommodation is itself a function of the animal's general genetic endowment—the workings of the biological machine—and so long as it ultimately allows individuals to grow into some shape that is capable of reproducing successfully, accommodation can be an adaptive response to a challenging world.

GUARDING DOGS AND SHEEP

Finally we turn to the role of accommodation in the behavioral development of livestock-guarding dogs. Early on in this book we posed the question of why the Maremma guarding dogs we observed in the Abruzzi Mountains would abandon their human shepherd in order to stay with the flock. Why did these dogs—apparently without instruction or command—"decide" to stick with sheep rather than a human? We offered up a raft of possible explanations. Perhaps they are genetically disposed to follow sheep and the Maremmas were displaying an intrinsic motor pattern for which guarding dogs were selected. Maybe they had been trained to act this way. Perhaps (though improbably) livestock-guarding dogs understand their job and realize that staying with the flock is important for avoiding predation. We wondered if it might be the case that any dog, not just a livestock guardian, would behave in this manner. But we don't think that any of those explanations is right. Rather, it's a story explained by accommodation to particular environmental factors during a critical period in the development of socialization.

After birth, dog pups remain with the mother and any sibling littermates until they're weaned from nursing and can eat solid food. This generally occurs in dogs by four to six weeks to even eight weeks into life. These are the critical weeks when they are developing connections in their sensory systems and brains that enable them to recognize social companions. As we've suggested, a kennel-raised pup that is separated from its mother at this juncture, away from its siblings and little-handled by humans, is going to be a spooky dog—shy of humans. It will shy away from domestic novelties, whether new objects in the environment or unfamiliar animals. What happens when a pup enters the critical period for socialization and is exposed to other species (us, or other dogs, or sheep)? Remember, Lorenz famously demonstrated that a greylag goose, under comparable conditions, can transfer social-bonding allegiance from its own mother to a member of a different species—himself. It doesn't even have to be a living animal, just an object that moves. Lorenz imprinted his geese on a water faucet hung on a string and moving at the right time. The same thing can happen with dogs. We and our students raised weaned livestock-guarding dog pups in pens they shared with lambs, "cross-fostered" by sheep ewes; they received very little human handling.

As adults, these dogs tend to be very shy of (and sometimes aggressive toward) people—but they are highly attentive to and gentle with sheep. Indeed, like Lorenz's geese, guarding dogs can be quite unselective in their bonding behavior during the critical period. The Maremmas we studied in the Abruzzi mountains of Italy generally preferred to attend to and follow sheep, and as we saw they will sometimes go off with sheep and abandon the shepherd. Interestingly, however, we observed some dogs that did neither: they balked at going out with their shepherd and stayed "home" when sheep were brought out to graze. Why? We concluded that these animals had actually bonded with milk cans.

FIG. 29 MOST MAMMALS AND BIRDS DISCOVER WHICH SPECIES THEY ARE DURING THE CRITICAL PERIOD FOR SOCIAL DEVELOPMENT. AS ADULTS, DOGS RAISED WITH LAMBS WILL TREAT SHEEP AS IF THEY WERE OTHER DOGS. PHOTO JAY LORENZ.

When they remained behind at the base camp, alongside milk cans and other articles of the sheep-milking business, they would stay close to the equipment and bark at approaching intruders just as they would if they were guarding sheep. Remember that much of the social bonding process in canids is done with the nose—and milk cans can smell as much like sheep as lambs do.

The bottom line is that if you raise a Maremma together with lambs it will pay close attention to sheep and stay close to the flock without disturbing them—and it won't have to be trained to do its job (fig. 29). The working behavior of a guarding dog is neither a product of learning nor an explicit genetic "blueprint." Nor were the individual behaviors (or lack of them) that make for a good livestock-guarding dog the target of intentional selective breeding. Rather, their useful behavioral shape arises as a result of accommodation between the dog's intrinsic motor-pattern repertoire and its early environment during a critical period in development. The overall accommodative shape—the right mix of intrinsic rules, the timing of development, and how those intrinsic properties respond to environmental inputs—can then be favored by selection, whether artificial or natural. Agriculturalists in sheep-raising cultures around the world have been producing good livestock-guarding dogs in just this way since time immemorial.

8 EMERGENT BEHAVIOR

Classical ethology, the early approach of Lorenz and Tinbergen, was rooted in a simple idea: much of an animal's behavior—the movement of its shape in space and time—is driven by intrinsic motor patterns, species-specific adaptations that are stereotyped products of natural selection and, as we like to say, critical parts of the shape of the "ticking mechanism" of the animal. A second idea is that some intrinsic patterns can undergo accommodative changes in shape that are due to developmental and environmental inputs. Taken together, we think these two ideas provide a pretty good general explanatory account of a lot of the behavior of dogs and other animals. But, like many scientific theories, it may be too simple. Indeed, there are many behavioral phenomena that aren't so easily explained in the elementary language of adaptation and motor patterns or in terms of accommodation of these intrinsic properties to external forces.

What is often called collective or collaborative or social hunting in wolves is a great illustration of a challenge to the simple ethological story. This is one of the most complex social behaviors in the carnivores: groups (packs) of animals appear to be able to intricately coordinate their activity in the pursuit of large prey such as moose, deer, or bison. When you watch a nature documentary of a wolf hunt, it's hard to avoid the impression that the hunting animals are cooperating, working closely together; that they know they're chasing the same deer; that they are keeping careful track of where their teammates are from moment to moment; that they can predict where the prey animal is going to run; and that they are capable of rapidly and intelligently adjusting and synchronizing their movements in order to cut off the prey's escape routes. Collective hunting therefore appears to be purposeful and insightful, requiring not a stereotyped pattern of intrinsic movements but rather a high degree of behavioral plasticity, organizational ability, and intelligence.

Our anthropomorphic tendencies are at the ready to reinforce this view: many of us would like to think that some animals, perhaps many, have a humanlike capacity to conceptualize a common goal, to share and communicate information, to make a plan, and to precisely coordinate the movements that are needed to carry it out. Consequently, many observers, behavioral scientists and nonscientists alike, are prone to conclude that this kind of highly complex behavior can't be just a simple result of the intrinsic predatory motor-pattern sequences that individual wolves might express when they forage on their own. Nor does it seem plausible to think that collective hunting—a social behavioral shape that changes from moment to moment—could simply be an accommodative remodeling of the intrinsic behavior of the many individuals that make up a group of hunting wolves.

Dog behavior poses similar challenges to a simple ethological approach. Some of the mundane and familiar things that dogs do—playing and barking, for instance—are puzzling from the standpoint of traditional ethology. Intrinsic behaviors are assumed to be adaptive—but when young pups play, it isn't at all clear that their exuberant behavior is functional in the ordinary adaptive sense of contributing to their ability to feed, avoid hazards, or reproduce. Play looks complicated and unpredictable (so much so that we devote all of the next chapter to it). In the same vein, we expect many species-typical motor activities, say, a vocal communicative signal like barking, to be stereotyped and predictable. But an individual dog's apparently "intrinsic" bark can in fact be extraordinarily variable in its acoustic shape, and barking can occur in a bewildering myriad of behavioral contexts. Simple ethological principles don't seem to be sufficient to make sense of this kind of complexity.

There is, however, an alternative and very different kind of explanation for how some complex behaviors might arise. In this chapter, we introduce a powerful third way of understanding behavior in animals — emergence. This is an old notion. Aristotle observed millennia ago that a whole is often more than the sum of its parts. That's certainly true of modern machines. An internal combustion engine has multiple components — chambers, valves, pistons, connecting rods, and driveshafts. These individual physical shapes are linked together, but, in and of themselves, they do nothing at all. Let an explosion occur in a combustion chamber, however, and the engine components suddenly come to life and conspire to produce motion. The true character of the machine is seen only when the parts interact: movement is an emergent property of the system that isn't inherent in any single part.

This venerable idea, in various forms and under many competing definitions, has gained a remarkable degree of intellectual traction in the last few decades in domains as diverse as physics, chemistry, biology, architecture, and human social psychology. The advent of computing machinery in the latter half of the twentieth century has strongly reinforced the notion. With a very simple physical and informational structure — essentially nothing but transistors that can be in an "on" or "off" state and a small set of basic instructions — extraordinarily complex outcomes are possible: solutions to thorny engineering problems, novel electronic music, rich animated imagery, even, some would say, a modicum of intelligence.

Granted, of course, that many of these remarkable products of computation are "directed" by human intelligence, by clever programmers who painstakingly manipulate the system's simple parts to yield a complex outcome. But some products of computing "machinery" are not directed and preplanned. You may be familiar, for example, with the Game of Life designed by mathematician John Conway. In this simulation of (something like) living organisms, the computer displays a grid of square cells. Each cell is either "alive" (perhaps colored red) or "dead" (uncolored). The cells interact with eight neighboring cells that are adjacent to them horizontally, vertically, or diagonally. At the start of the game, the cells can be filled randomly. The game then proceeds in steps. At each step the cells obey a few simple rules:

1 If a cell has fewer than two live neighbors, it dies.
2 If a cell has two or three live neighbors, it lives on to the next step.
3 If a cell has more than three live neighbors, it dies.
4 If a dead or empty cell has exactly three live neighbors, it becomes a live cell.

When this game runs step by step on the computer, following only these four rules operating on some given initial state, an extraordinary variety of complex patterns of colored grid cells begins to emerge: shapes change with each step and they exhibit (apparent) movement. You can find lots of great demonstrations of this effect on the web, as well as little programs that will allow you to run this "cellular automaton" yourself. The point is that none of the actual resulting patterns (except the initial state of filled cells) were preprogrammed or defined in the software or "wiring" of the machine. They are emergent products of the system.

One definition of "emergence" (and there are many) is that complex and novel properties can come about by the undirected "self-organizing" interaction of simple rules and processes. Thoughtful scientists, philosophers, and mathematicians have been intrigued by the possibility that emergence might ultimately provide answers—perhaps unexpected—to some of the very hardest problems in science: What does it mean for an organism to be "alive?" How does the complex organization of the human brain result in consciousness and intelligence? How does human language develop and function? Why does the cosmos itself have the properties that it does? Our questions about dog and wolf behavior—cooperative hunting, complicated parental behaviors, barking, and social play—are rather more modest, of course. Nevertheless, we think that the notion of emergence has the potential to shed some light on how biological form and behavior arise by means other than direct selection for intrinsic shape or the reshaping of structure and behavior by accommodation.

To get a feel for what an explanation by emergence in the biological world is like, let's begin by thinking about termites. These insect invertebrates are, of course, much simpler than a wolf or dog—but paradoxically they exhibit an almost uncanny degree of behavioral complexity.

FIG. 30   A COMPLEX TERMITE MOUND IS CONSTRUCTED BY THE INTERACTION OF A
COUPLE OF EXTRAORDINARILY SIMPLE BEHAVIORAL RULES. PHOTO BY DANIEL STEWART.

Consider their elaborate mound structures, the "castles of sand" that are common sights in arid deserts and plains (fig. 30).

Termite mounds are vastly bigger than the individuals that build and inhabit them. They have massive walls, soaring chimneylike spires, and multiple entrances, exits, and internal chambers. And the mounds, which are typically found in extremely hot dry climates, feature highly efficient cooling systems that can maintain a constant temperature in the heat of the day and the cold of a desert night. Could this kind of intricate design and engineering sophistication possibly arise from the genetic programming of intrinsic behavioral properties in individual termites? Could it be that groups of termites collectively have a high order of conceptual capacity, intelligence, and ability to manipulate their physical environment? Neither of these explanations seems plausible. It is clear that individual termites don't have specialized genes that are dedicated to the complexities of mound building and design. Nor do they have body shapes or brains that can explain their complex building skills. So how do they manage to do it?

The best bet is that the moment-to-moment activity of mound building actually consists of a small set of basic and simple motor patterns. Termites pick up grains of sand with their jaws, carry them somewhere, and drop them; they sense and tend to move toward other termites; and their movement is sensitive to environmental conditions such as humidity and concentrations of gases in the air. No single termite does much more than that. However, the interaction of many, many thousands of termite workers over long periods—each doing little more than picking up a sand particle, moving along with others in response to particular environmental properties, and then depositing their load—eventually conspires to generate a structure that couldn't be planned or built by a single individual. The complexity of termite building behavior certainly isn't encoded in the genes of these very simple organisms—it isn't caused by natural selection yielding specialized engineering abilities. It isn't learned and it isn't planned. It doesn't result from the exigencies of developmental accommodation or from learning or intelligence. It appears simply—but remarkably—to be a consequence of the emergent interactions of some few basic behavioral rules.

It is likely that that many remarkable biological and physical phenomena are best understood in this way. Before returning to behavior, let's stop for a moment to appreciate a few such examples. A coiled snail shell, for instance, offers a lovely illustration of the power of emergence in the determination of physical form. At first glance, this beautiful kind of structure appears to be remarkably complicated. But look at one more closely and you will see that the initial innermost shape is itself quite simple: it is a single small tight spiral. This is the earliest form of the shell in the course of a snail's development. As the snail grows, it adds onto the initial shell structure, increasing the radius of the spiral exponentially: first it doubles in size, then quadruples, and so on. So long as the "angle of departure" of the spiral remains constant, the shell will have a successively larger but always self-similar shape. This means that, at any scale, the elegant design has the same properties—it simply gets larger and larger according to exactly the same basic "rules" of growth.

The snail shell's complex shape is a result of intrinsic characteristics of the cells that produce shell material. The remarkable overall form that emerges, however, isn't itself a preprogrammed intrinsic property of the snail genome. Nor is it an accommodation to environmental forces acting on the shell (though external factors can also add to the developmental mix and produce an astonishing range of different shapes in different species, which may also have different intrinsic growth rules). The apparent complexity and beauty of the coiled snail shell is an emergent effect of the interaction of simple rules of growth.

The organization of much more complicated organs—for instance, the mammalian brain—can also be understood to result, at least in part, from emergence. In a remarkable recent study by a research group at the Institute of Molecular Biotechnology in Vienna, Austria, it was shown that when certain fated human embryonic cells are cultured by a technique that provides structural support for growing tissue, it is possible to grow a three-dimensional "mini-brain" in the laboratory. The intrinsically fated cells are destined to become neurons. What is remarkable is that these growing neural cells self-organize into particular and distinct brain shapes, much like the specific well-defined regions that characterize the gross structure and organization of a normal developing brain. The basic growth rules by which these neural cells proliferate (in a "dish") interact to produce complex emergent structures.

GEESE IN A V

If the shape of complex organs can arise by emergence, it follows that behavior can, too—because behavior is the shape of animals moving in time and space. Think, for instance, about the Canada geese we often hear up above us in the fall, noisily honking away. They are social animals that fly together in flocks. During biannual migrations their flight takes on the famous V shape, as shown in figure 31.

The first question to ask is whether this behavior is intrinsic. Are there genes that somehow determine the wedge shape of the flight pattern in geese? Is the V pattern a product of natural selection? It seems pretty farfetched to think that geese that fly in V patterns leave more offspring than those that don't. Nevertheless, flying in Vs is certainly a species-typical taxonomic characteristic of Canada geese. Every time you hear them up above in the fall, honking away, they are flying in

FIG. 31 MIGRATING GEESE IN A V FORMATION ARE ALSO FOLLOWING SIMPLE RULES.
DRAWING BY JOHANNA AULÉN.

some form of a V—they all do it. That suggests some kind of genetic basis for the behavior. In fact there are many species of social birds that fly in the V or wedge pattern—brants and cormorants, for instance, typically fly this way as well when they migrate. When several species of birds exhibit the pattern, we might conclude that, like the homologous inheritance of the EYE > STALK motor patterns of the carnivores, it was inherited from some ancient and common ancestor.

But here are some telling differences between these behaviors. Unlike the EYE > STALK that each animal can perform alone, flying in a wedge pattern is a social behavior sometimes involving hundreds of birds. You can't have a "WEDGE" motor pattern with only one bird, but a single Border collie can perform EYE > STALK. Additionally, it is pretty easy to see how the EYE > STALK motor pattern confers a selective advantage. Animals that assume the hunting shape had (and still have) a better chance for survival. If a predator detects prey off in the distance and simply runs at them headlong, the prey is likely to spook and escape. An animal that just breaks into a run is going to be less successful than a predator that sneaks up on the prey, getting close to it before crossing the threshold and going into a surprise attack with a short chase.

From the standpoint of energy expenditure, the short chase is most efficient, and so you can imagine how the threshold distance between STALK and CHASE could evolve. Simply put, those animals that break into CHASE too far away from the prey put more energy into pursuit and are likely to be less successful in catching it. Those that keep sneaking up and get too close are more likely to be noticed, and the surprise is gone. Cheetahs go into EYE > STALK when hunting animals like gazelles — but if the gazelle looks at the cheetah before the predator crosses its threshold for CHASE, the cheetah ends the predatory sequence. You can see the selective advantage in that. The likelihood of a successful hunt goes down if the cheetah is discovered. If the cheetah chases the gazelle and doesn't catch the prey, it will have spent the energy with no benefit.

When we try to apply the same adaptive logic to our wedge of geese, we have a problem. With the cheetah, selection operates on each individual animal, for or against its behavioral shapes and the thresholds between them. Too many failures with a particular shape, and that animal will starve and fail to leave offspring. With the social geese, we have to ask how each individual goose might gain a selective advantage by flying in a wedge. Is it beneficial for foraging, reproduction, or hazard avoidance?

Group behavior certainly can seem to confer a selective advantage. Many fish form schools — a collective shape that offers a better chance of avoiding predation. When threatened by predators, the rule is to swim toward other fish (of the same species) and form a large swirling group where it is very hard to target a single fish. All they have to do to form a school is swim toward another fish. Birds do that too. They avoid predation on occasion by flying in huge flocks, where a hawk would have a devilish time flying into the swarm and trying to target one bird. So being social can constitute a hazard-avoidance activity. It gives each individual animal a better chance of surviving an attack. But what about the V? Does that particular shape deter or confound predators?

In other cases, group behavior enhances foraging ability. Many terns come together in a feeding frenzy when one individual observes another that has discovered a school of fish. In effect, the finder has special information and signals to others that food has been found. Maybe our wedge of geese flying south in winter are also following some wise individual who knows where winter food is to be found. But why do it in a V?

In the same vein, you might argue that goose social behavior confers a reproductive benefit. If flocks are family groups where chicks are following their parents south to winter feeding grounds, perhaps this is a form of parental care with a selective advantage for both parents and offspring. But why a V?

Maybe we are approaching the V formation the wrong way. Let's think about what the geese are actually doing. They are flying south, in the fall, because the days are getting short. Is this a social behavior? No. The crucial experiment has been done many times on many species of birds. Put your subjects in a cage under artificial light and start reducing the day length. At some point they begin to move actively toward the south side of the cage. This is a simple intrinsic rule: when the days get short enough, fly south. Millions of birds do this every year all by themselves. You don't need to be in a flock to do it.

So if robins and thrushes all fly south by themselves, why do geese flock in a V? Remember that Canada geese are big animals—a large male can weigh twenty pounds. Few species of flying birds ever get over twenty-five: it just isn't possible to have wings and muscles strong enough to lift twenty-five pounds off the ground and a metabolic "engine" that will then allow such a large animal to fly by wing flapping alone for thousands of miles. Consequently, many of the bigger migrating species are gliders and soarers. That may account for why some South American condors are so large: they live in a mountainous habitat where there are frequent powerful updrafts that can supply the energy to sustain flight. But the niche of the Canada goose extends from the far north in summer to mid-Atlantic regions in winter. It would be hard to find enough updrafts on the thousand or more miles between the Arctic and the Chesapeake Bay to glide the whole way.

So the migratory trip south is going to be difficult for big geese. Given their size, they can't beat their wings for the whole time; they need some kind of more efficient propulsion. One method of easier travel is drafting—that is, following in the wake of another goose. Drafting can be very efficient indeed. One of us drafted his Ford van behind a trailer truck during a long drive on the highway. The van normally got between eighteen and twenty miles per gallon. With drafting, it got twenty-six miles per gallon. That was a 25 percent saving in energy. For a migrating goose, the best method is to draft off the big goose ahead of you, which, because of its size, is disturbing a lot of air. As the goose in front flaps its wings it causes air on each side to go up and down in pressure. Each time the pressure reduces, the air will stream back in to fill the vacuum. One goose following another can ride on the bubbles of air that are created behind (and slightly higher) than each of the leader's wings — on one side or the other of its body.

To take advantage of this effect, the follower can't be directly behind the leading goose. So if there are two geese drafting off a leader, each will be to one side of it or the other. That is the start of the wedge formation. Behind the leader and the first follower, two more geese can draft slightly to the side and behind them; this pattern recurs as far back as there are geese to ride the bubbles. What we observe is a group of birds flying in some sort of V. On this view none of them has a V "in mind"; that particular shape isn't "in the genes." Rather, each individual goose follows two simple rules: (1) fly south and (2) follow in the wake of another goose (if there is one).

The first goose to take flight, following rule 1, is the leader (and on any individual flight will have to expend more energy; we hope they take turns). Every other individual obeys rules 1 and 2. The apparently complex and "special" V flight pattern is simply an emergent effect of the interaction of these intrinsic adaptive rules, an artifact of flying the easiest way possible.

Yes, the flock of geese looks like a social group in which individuals have a special association or affiliation with one another. But that isn't necessarily the case. It is possible that none of the geese in the flock "know" each other at all. All each goose has to do is find a place in the environment where it is easier to fly south. That place just happens to be behind another goose — and any goose will do.

Ethologists investigating some pattern of apparently organized collective activity often assume it must be an adaptive social arrangement. Pair bonding in wolves, coyotes, or jackals is a good example. Whether you look at the scientific literature or watch popular TV nature documentaries, you'll find the assumption that animals like wolves bond together in pairs that may last for a lifetime. It sounds idyllic, like a happy marriage. For some, it suggests evidence of intelligence.

An alternative view is that pair-bonding behavior (at least in carnivore predators like the canids—the story is perhaps more complicated in many bird species) is an emergent character that results from the interaction of two simple behavioral rules: (1) if you are a female, protect your feeding territory from other females, and (2) if you are a male, defend your feeding territory from other males.

If a male and a female happen to express these foraging rules in the same territory, and persist in that territory over time, they are most likely to mate with one another. We say "most likely" because if either of them moves close enough to the territorial boundary, they may mate with some other animal. What tends to prevent promiscuity in these circumstances is that the underlying simple behavior for both sexes is, in fact, to ward off same-sex competitors. The apparent pair bonding is an emergent artifact of this territorial defense behavior.

Dogs, in contrast, don't have feeding territories. They are scavengers who follow piles of food wherever they are available. It's true that dogs will growl and snarl to dissuade competitors at a food bowl or in a backyard where a household dog might be routinely fed. But this food-defense behavior in pet animals is limited to a very small area. Human garbage dumps are also great places for less-restricted animals to scavenge. In the Mexico City dump we found some seven hundred dogs living and feeding in one of the world's largest landfills, some nine hundred acres. In that case, where food is plentifully distributed over a large area, we don't find territorial food-defense behavior (and it is rare in pet dogs beyond a very limited range). And, in fact, both male and female dogs are totally promiscuous. Without large-area feeding defense, potential mates are freely available. That is why breeders famously need to carefully protect females and often send them miles away when they want to get them bred to a specific male. In less-controlled circumstances (for instance, among sled dogs), it is often very obvious that there were two or three fathers to a litter. The absence of pair bonding in dogs can thus be understood as a "negative" emergent consequence of their feeding and territory-defense behavior: dogs lack the simple rules whose interaction would otherwise cause pair bonding to arise.

We see a similar story in the parental feeding behavior of wolves, coyotes, and jackals, animals that have some of the most sophisticated parental care in the carnivore world. Most of the carnivores (though not dogs) will provision their young. This simply means that they kill something and bring it to the offspring or lead them to a kill. The wild members of the genus *Canis* have an additional method of feeding their young: the regurgitation of partly digested food. REGURGITATE is a motor pattern where a specific begging performance by pups elicits the vomiting behavior (plate 5). It can in fact, be elicited from any adult. Thus last year's pups (now a year old) can be recruited to feed this year's young. Nonparent wolves in captivity—on the same "territory"—will indeed regurgitate to pups. We don't often see REGURGITATE in non-relatives, however—because they don't tend to be allowed on the feeding territory. So we can understand the wild-type provisioning behavior as an emergent consequence of two rules: (1) stay on a territory, and (2) REGURGITATE.

The rules must interact if provisioning behavior is to appear. Domestic dogs do have the second rule—they occasionally (but very rarely) regurgitate food for their offspring. But they don't have the territorial rule. In effect, like the absence of pair bonding, the fact that dogs don't provision is simply a "failure" of emergence, that is, the requisite rules that would otherwise produce the emergent effect are not all present.

"COOPERATIVE HUNTING"

Yet another form of canid foraging behavior that might best be seen as a product of emergence is cooperative hunting in wolves. This phenomenon, as we've noted, is often regarded as an example of a sophis-

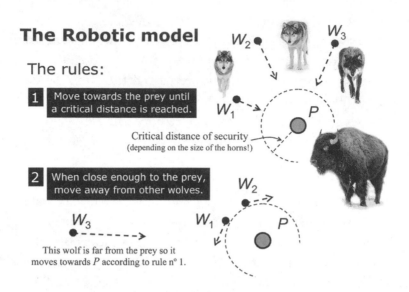

The Robotic model

The rules:

1 Move towards the prey until a critical distance is reached.

Critical distance of security
(depending on the size of the horns!)

2 When close enough to the prey, move away from other wolves.

This wolf is far from the prey so it moves towards P according to rule n° 1.

FIG. 32   TWO SIMPLE BEHAVIORAL RULES GIVE RISE TO WHAT SEEMS LIKE COMPLEX "COOPERATIVE HUNTING" IN WOLVES. DIAGRAM BY CRISTINA MURO.

ticated and highly complex adaptive behavior or, indeed, as a remarkable instance of animal intelligence. We're not so sure. Ray and several computer scientist colleagues (C. Muro, R. Escobedo, and L. Spector) designed and implemented a computational model of collective hunting behavior to see what would happen when wolves and prey were represented by abstract "robots," digital agents that were programmed to do no more than carry out two simple virtual motor patterns (fig. 32).

The rules they expressed were both local and decentralized: an individual agent (a "wolf") could act, or not, completely independently of the behavior of any other agent, and there was no coordinated signaling between participants. These rules were: (1) move toward the prey until a specified safe distance is reached, and (2) when close enough to the prey, move away from other wolves.

The outcome of this simulation was a successful pattern of prey capture that looked remarkably like real collective hunting behavior. The virtual animals appeared to be moving in concert, adjusting their relative positions and circling around prey. Eventually, one simulated wolf would find itself close enough so it could capture the target.

Nothing in this model depended on the presence of intelligence,

FIG. 33 IF WOLVES OBEY THE TWO RULES IN FIGURE 32, THEY END UP CIRCLING THE PREY
EVERY TIME. PHOTO NATIONAL PARK SERVICE/DOUGLAS SMITH.

purposive intentional behavior, communication, or a hierarchical social structure (where, for instance, dominant "alpha" wolves might play
leading or guiding roles). Rather, the pattern of complexity emerged
from the action of individuals following simple interacting behavioral
rules (fig. 33).

The same kind of result has been found by other researchers using
virtual swarms of simulated animals (and even physical robots) to reproduce complex group behavior in birds, fish, ants, and many other
species. Sergio Pellis and Heather Bell, for instance, studied huddling behavior in newborn rat pups. "Imagine," they wrote, "a litter
of 7 day-old rat pups being placed, scattered, on a tabletop. The table
has a smooth, flat surface and walls on its four sides. After some time
elapses, the pups huddle in a pile in one of the corners. The problem is
to explain how and why this aggregation occurs."

Pellis and Bell note that young rat pups aren't capable of maintain-
ing their body temperature. By aggregating into groups they reduce the
speed of heat loss. But how does the actual shape of these groups arise?

Pellis and Bell suggest that just two simple behavioral rules are required to explain the huddling pattern: rat pups prefer to move toward vertical surfaces, and (like dog pups) they are thermotaxic, preferring warmer to cooler surfaces.

"When placed on the tabletop," they observe, the pups "begin to move, and when they encounter vertical surfaces they remain in contact, hence the aggregates typically form against a wall, and usually in a corner (i.e., greater vertical surface area). However, once another pup is close, it is warmer than the wall, so the pups are attracted to each other rather than the wall. Consequently, over time, the pups aggregate in one of the corners."

Interestingly, this picture changes with age. By the time the rat pups are ten days old, they tend to form groups only with more active littermates—sluggish individuals, it appears, get less of the heat-saving benefit of huddling. This difference, Pellis and Bell conclude, results from a developmental change in the perceptual world of young rats—they become more sensitive to movement—and this can be described by the addition of a third simple behavioral rule: move toward more active littermates. Several researchers carried out computer simulations that utilized these rules, and successfully modeled rat huddling with abstract agents. Nothing more needed to be said about the rat pups' emotional state, intentions, or anything else—two simple rules alone generated the grouping behavior, and adding a third rule into the simulation, to model development, changed it appropriately.

What about dogs? In a recent study, a group of British and Swedish researchers showed that shepherding behavior in Border collies can be simulated as well by a self-organizing computational model in which the herding dog follows two elementary rules: round up sheep with the "algorithmic" equivalent of the Border collie's breed-typical EYE > STALK > CHASE motor pattern, then CHASE them forward as a group. The simulated sheep likewise obey just two rules that are in fact typical of the intrinsic motor patterns of many social herbivores that graze as a group: move toward one's nearest neighbor, and move away from a potential threat. This model—which could readily be implemented with robotic machines—does a fine job of emulating the movement of (actually observed and recorded) collies at work and in herding trials, an apparently remarkable feat that is often attributed to their "innate intelligence" and the ingenuity of human trainers. But this study suggests that the overall form of these complex patterns of behavior can be explained just as readily (and much more simply) by the principles of self-organizing emergence.

BARKING EMERGES

Barking is likely to be the very first thing you think of when asked, "What do dogs do?" It is a hallmark characteristic of the animal—it seems like it might be a great example of Lorenz's dictum that a behavior is a defining taxonomic trait of a species. It's hard to find any place in the world, at least near human habitation, where you won't be beset at some point by loudly barking dogs. It occurs so frequently in the behavioral repertoire of many animals that it is often described as hypertrophied (the opposite of atrophied). All dogs bark, though the frequency varies in different breeds, and they begin to do so very early on in life (at about two weeks). Moreover, as we noted earlier, we raised a litter of congenitally deaf Border collies that barked normally in spite of a total absence of auditory experience. In all these respects, the bark seems like an obvious candidate to be listed as yet another intrinsic motor pattern in the ethogram of the dog.

But motor patterns are, by definition, stereotypical—they have the same general characteristics in every member of a species. Dog barks are anything but that. In one animal, the vocalization is loud and noisy, with sound energy at many random frequencies, while another dog's bark sounds tonal and "purer." A dog may bark just once (though that's rare in dogs—single-pulse barking is much more common in wolves), or almost unceasingly. We once listened to a livestock-guarding dog in a field bark for nine hours until it became so hoarse that it went through the articulatory motions but could no longer actually produce any sound.

Indeed, the same dog can exhibit barking behavior with any or all of these characteristics. Moreover, the classical idea is that motor patterns are triggered by specific releasers. For example, the LOST CALL is triggered by a temperature differential, and itself triggers RETRIEVE.

Dog barking can seemingly be set off by just about anything at all—a rabbit running across the lawn, the moon at night, leaves crackling, anticipating a walk with its master, the approach of a stranger . . . and certainly by the barking of other dogs. So even though it appears to be an intrinsic, ubiquitous, and frequently expressed species characteristic, and it certainly has a recognizable quality (or range of qualities), the bark actually doesn't fit the definition of a classical simple motor pattern especially well.

Zoologist E. S. Morton, long-time research scientist at the Smithsonian's National Zoological Park, has observed that many mammalian (and avian) vocal motor patterns seem to reflect a small set of rules. Low-frequency and noisy sounds like a growl, Morton observed, are common in the vocal repertoire of many mammals—including dogs—and tend to occur when the sender is disposed to act aggressively and "wants" the hearer to withdraw: "Watch yourself! Back off!" By contrast, high-pitched tonal signals like a pup's whimper (or the LOST CALL) generally tend to be associated with appeasement and care solicitation. They indicate that the signaler is not a threat. It's OK to approach: "Come closer! Take care of me! Stay with me!" Both of these vocalization types occur in dogs and related canids. You can think of them as two fundamental and simple rules of vocal behavior that reflect intrinsic properties of the class of mammals.

We once observed a Portuguese Estrela Mountain dog defending a flock of goats. It vocalized when we approached. The call sounded something like a bark—remember, what we call "barks" are highly variable—but it was very brief, not particularly loud, and entirely nontonal. You might describe it impressionistically as a soft, noisy huffing sound; we'll call it a WOOF. It is an example of a signal that follows the first of "Morton's rules." And we see exactly the same WOOF motor pattern in wolves, coyotes, and jackals. It is a signal that indicates to the hearer that it ought to get lost quickly—an intrinsic hazard-avoidance motor pattern.

One twilight evening in northern Minnesota we watched five wolves coming to feed at a garbage dump. We thought we were fairly well hidden, observing quietly and unobtrusively, until one individual detected

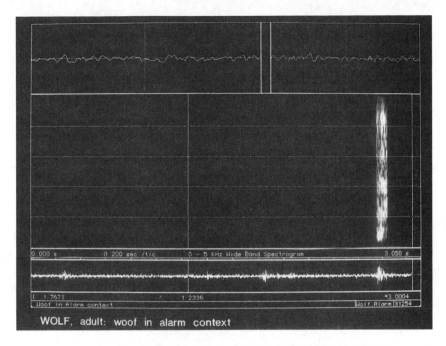

WOLF, adult: woof in alarm context

FIG. 34 SONOGRAM OF A WOLF ALARM CALL, A SHORT, NOISY SIGNAL. NOISE IS ACOUSTIC ENERGY AT RANDOMLY RELATED FREQUENCIES, AND THERE IS ENERGY AT JUST ABOUT EVERY FREQUENCY (*REPRESENTED ON THE VERTICAL AXIS*) IN THIS BRIEF WOOF.

us and emitted that single low noisy WOOF (fig. 34). We were startled ourselves. On hearing the call, the wolves instantly vaporized away into the night, a consequence of Morton's first rule in action.

By contrast, a pack of beagles on a hunt follow the second of Morton's rules. When they detect a rabbit (or its olfactory trail) they produce a characteristic loud, high-pitched, and tonal cry—the "scent bark." The outcome of giving this tonal signal is that the whole pack, along with its companion human hunters, rallies together as they go off on the chase (fig. 35).

These two canid vocalizations are instances of each of Morton's rules, and both are classical intrinsic motor patterns. They are stereotyped, they aren't learned (a beagle can't be trained not to give the scent bark), and they are released by a very particular kind of environmental input. Different dog breeds display a wide range of these TONAL

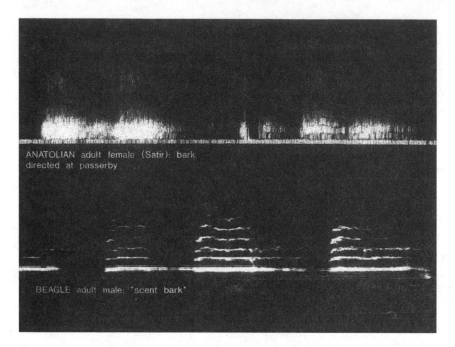

FIG. 35 THE TOP VOCALIZATION IN THIS SONOGRAM IS OF AN ANATOLIAN SHEPHERD
NOISILY BARKING AT A STRANGER: IT SAYS, IN EFFECT, "GO AWAY, GO AWAY, GO AWAY . . ."
THE LOWER TONAL BARK IS A BEAGLE THAT HAS JUST FOUND THE TRAIL—SIGNALING
"COME HERE, COME HERE, COME HERE."

and NOISY motor patterns—whines, whimpers and cries, along with growls, grunts and woofs. They are all part of the ethogram, the basic behavioral repertoire of all dogs (indeed, all of the canids).

Now consider the Maremma guard dog pictured in figure 36, chained to a tree. When we describe its overall behavioral shape, we observe that the dog's head is lowered below the line of its back. Its mouth is opened to reveal a hint of teeth. The hair on its back is raised (piloerection). The dog's legs are stiff. The tail is held low and almost between the legs. People who "know dogs" are likely say that this animal looks aggressive. (We often use the term "aggression" as though it were a unitary thing. But it isn't—it is a complex combination of shapes, and indeed we might well think of aggression itself as an emergent effect.)

Because this dog was chained to a tree it could not flee the intruder, and it could not fight. Nor was it able to move and change its shape in response to a threat. Here was an animal in conflict: both flight and

FIG. 36 THIS MAREMMA IS TIED TO A TREE AND BEING APPROACHED BY A STRANGER WITH A CAMERA. THE DOG CAN NEITHER ATTACK NOR RETREAT. IT'S IN CONFLICT AND CAN'T RESPOND APPROPRIATELY. THE DOG'S HEAD IS DOWN, AS IS THE TAIL, WHILE ITS HACKLES ARE UP AND IT IS SHOWING TEETH AND BARKING. IS IT AGGRESSIVE OR SUBMISSIVE OR BOTH?

fight would have been appropriate responses but neither was possible. What it did do was bark. Why?

Faced with an intruder, we might expect the dog to follow Morton's first rule and emit a noisy "aggressive" signal like a WOOF. But unable to act independently, the dog is also disposed to express Morton's second rule. We would, that is, expect it to produce a tonal motor pattern (like the rallying call of a beagle or the insistent crying of a pet dog left alone) that could appease the intruder or call in conspecific reinforcements. Faced with this conflicting state of affairs our Maremma emitted a signal that was a composite of two elementary vocal motor patterns—part TONAL and part NOISY. And in fact, the vast majority of dog barks have exactly this mixed acoustic character. You can see it clearly in the spectrogram in figure 37, which was made from a recording of the dog in the photo above.

Like the chained Maremma, household dogs are often restricted in

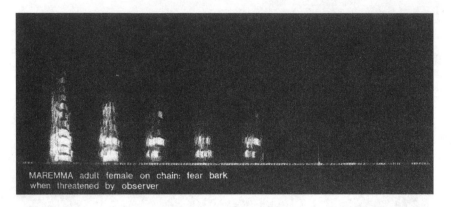

MAREMMA adult female on chain: fear bark
when threatened by observer

FIG. 37 SONOGRAM OF A BARK PRODUCED BY THE DOG IN FIGURE 36. THIS VOCALIZATION
HAS TONAL ELEMENTS THAT SIGNAL THE RECEIVER TO APPROACH. IT ALSO HAS NOISY
COMPONENTS THAT INSTRUCT THE RECEIVER TO GO AWAY. THE CONFLICTED ANIMAL
SENDS A SIGNAL THAT SAYS, IN EFFECT, "COME HERE, GO AWAY."

their movement. They are constrained by leashes, fences, apartment walls and doors, or even the commands of human owners. Consequently, these animals are likely to be as motivationally conflicted as the chained guarding dog—and they do bark a lot. Unfamiliar intruders in odd clothing—like the proverbial mail carrier—routinely violate a dog's space while it is tied up or otherwise limited in action. It's no wonder mail carriers get barked at so often (if not bitten when the dog manages to escape its constraints).

Wild canids can also find themselves in these kinds of conflict situations, particularly in captivity. An Oregon farmer, for instance, asked for one of our livestock-guarding dogs to protect his sheep from a wolf that he kept in his yard—perhaps unwisely, as it periodically escaped from its chain to threaten the flock. When we observed the wolf, it was restrained. As we approached the chained wolf, it ran round and round barking exactly like a dog. This was clearly an animal in conflict—it couldn't attack us and it couldn't flee. In nature, by contrast, you rarely hear a wolf bark: Ronald Schassburger has reported that barking constitutes just over 2 percent of total vocal behavior in wolves, compared with 96 percent in dogs. But put one behind a fence at a zoo or at a wolf reserve and you will hear it bark far more often.

As we've emphasized, dog barking isn't stereotyped. Some barks are highly tonal, and sometimes they're quite noisy. Often they're repeated over and over and over, with differing degrees of mixed tonality and noise. Why this variability? It depends on just how much and how long the animal is restricted—and how strong the motivational conflict may be. Pet and companion dogs are frequently confronted by a myriad of conflicting human demands and environmental pressures. In contrast, unrestrained household dogs, or village dogs on the street who move freely on their own, find themselves in a state of conflict far less often. Like wild canids (when they're not fenced in), free-living dogs bark much less frequently and with far less variability than familiar household pets.

How, then, should we understand what a "bark" really is? As we see it (and as did Morton, who talked about bark-like vocalizations in the same way in his original seminal article on the subject), barking is not a unitary type of vocal signal. It is not a single motor pattern. Rather, it arises when a conflicted animal is motivated to simultaneously produce two basic intrinsic motor patterns, each expressing a simple rule:

1. Produce a NOISY vocalization (i.e., vibrate your focal folds in an irregular aperiodic manner) when you're motivated to signal potential aggression or to encourage the withdrawal of an intruder or a threat.
2. Produce a TONAL call (i.e., vibrate your vocal fold regularly and periodically) when you're motivated to appease an aggressor or to draw the attention of potential helpers to deal with a threat.

The variability of barking depends on a particular dog's internal state and the contingencies of what is happening in the world around it—it is a function of which NOISY and TONAL rules happen to be engaged by an animal in conflict, and to what degree. One conflicted dog may be disposed to WOOF in alarm and to WHIMPER to solicit care, another, to briefly GROWL and WHINE. Each combination will result in a somewhat different pattern of acoustic energy, of noise plus tonality. What we like to call the bark isn't a single motor pattern in its own right; in spite of the fact that it is so typical of dogs, it isn't an intrinsic characteristic of the species. Rather, we think that barking is another great example of emergence: a complex result of a few simple but interacting behavioral rules.

9 PLAY

Hundreds of scientific papers have been written on the subject of "play" behavior—an activity for which dogs are, of course, famous. Their playfulness is one of the reasons that so many of us take them into our homes and love them. You can find many examples of the vast literature on animal play, some old and some very recent, in our bibliography. But getting a handle on what play really is, and what role it serves in the lives of animals, has long been a challenge. One of our favorite writers on the subject, Gordon Burghardt, has been working on the problem since the 1980s. In an article thirty years later, he still describes play as "an enigmatic behavior that is hard to define."

Just about every account of play starts by acknowledging that it is one of the great mysteries in the evolution of behavior. When ethologists say that a behavior has evolved, we mean that, like an intrinsic foraging motor pattern or the capacity of an organism to accommodate to its environment, it is in some way passed on through the genes; animals that engage in play must have had a selective advantage over those that didn't. We humans usually think we know play when we see it, and often it seems to have a special character all of its own— it looks like "fun." But is having fun a sufficient reward to shape the biological evolution of a behavior? Most ethologists who have thought about the problem agree with Burghardt, that play is "seemingly nonfunctional"—it doesn't play a primary role in supporting the basic requirements of an animal to feed, avoid hazards, and reproduce. In the previous chapter we argued that complex types of behavior may arise by emergence rather than as direct products of natural selection. Here we will try to show that—like pack hunting, the flight formation of geese, or barking in dogs—play behavior is best understood as an emergent phenomenon.

THE NATURE OF PLAY

Ray and his wife Lorna once had a Jack Russell terrier that would zoom in and out of its doggie door like a madman, skidding on the hardwood floors, only to repeat the performance over and over. It didn't do it for a treat or a pat on the head—but, rather, to all appearances, out of sheer exuberant joy. It looked for all the world like a playing child trying to impress with its athletic ability. And outside the window as we write this, we're watching a pair of yearling white-tailed deer in a field—apparently frolicking with a wild turkey. These herbivores aren't going to catch and eat the turkey. It's certainly no threat to them. And it's hard to imagine how engaging in this kind of "playful" behavior could possibly help them become better adults. They just seem to be stalking the turkey because it's there and they have nothing else to do. The deer do look like delighted kids at play. The turkey, who is trying to feed at a pile of seed left out for it, doesn't seem amused.

Of course, if you want plenty of evidence for this sort of "play" behavior you don't need go any further than your own or your neighbor's pet dogs. They are likely to be chasing balls, tossing Frisbees with enthusiasm, or nipping frantically at each other's tails. We are usually sure that we know play when we see it—so we will assume that readers know what we mean by the "behavioral shape" of play. We'll also assume that when we talk about animals playing we are referring to mammals. Birds and reptiles are occasionally described as engaging in play, and recent studies have observed play-like behavior in invertebrates like the octopus and even in spiders. But our advice is that if you really want to understand play, you have to start with a mammal like the dog. You'll certainly get a lot more usable data a lot more quickly.

And you have to make sure you start with juvenile mammals. Forget neonates—they're about as humorless as crocodiles. There's little doubt, however, that play is especially common in young developing animals. It's true that some species exhibit play in adulthood—adult dogs certainly do and so, of course, do some humans—but the frequency of play in later life has not been well-studied. It's sometimes hypothesized that dogs continue to play throughout their lives as a way of reinforcing the social bonds of domestication. But we've seen great examples of adult wolves apparently playing too—have a look at plate 6. The common perception that adult dogs play a great deal more than their wild counterparts, or more than other adult animals in general, is probably biased by the fact that we are surrounded by millions of them. All in all, however, a young dog is a better study animal than an adult rhinoceros.

What more has to be said, then, other than that play is just another especially interesting behavior in the dog ethogram? Couldn't play, even in its great variety, simply be a product of natural selection? As we said, most ethological accounts of play take it for granted that this is the right way to think about it—as an evolved intrinsic property of animals, a genetic disposition that is, perhaps, also shaped by accommodation in the course of development.

The story is not quite so simple, however. "Play" in dogs (and other mammals), as we've said, is a very strange phenomenon from an ethological standpoint. People seem to have no trouble identifying it when they see it—but precisely characterizing what a dog or other animal is actually doing when it plays is a great challenge to the ethologist. That's why we often put the word in "scare quotes": in spite of the fact that people feel like they know it when they see it, it's not at all obvious that play is a unitary "thing-in-itself" that can easily be characterized, let alone explained in evolutionary terms.

Play behavior comes in many forms. In a recent review, J. W. S. Bradshaw and colleagues at the School of Veterinary Sciences of the University of Bristol observe that play displayed by solitary individuals is typically directed at objects like balls or chew toys (in dogs and other canid predators these "play objects" are often preylike and are acted on as if they were prey). But this kind of object play also appears in social contexts—famously, of course, between dogs and their human owners—in what look like "games" of competition between two or more individuals, tugs-of-war over possession of some object. In addition, play occurs without being directed at inanimate objects, for example, in aggressive ("agonistic") bouts of mock fighting between two individuals or in extended chasing of one animal by another (plate 7). Even a single dog chasing its own tail might be said to be at play.

Indeed, the charm of watching animals at play is that their activity is often so wonderfully unpredictable—sometimes so utterly unexpected. Nature documentaries and YouTube videos routinely showcase examples of what looks like remarkably inventive play: young otters joyfully slide in the snow; polar bear cubs wrestle; cats tap away at piano keys. All these animals seem to be doing something that would generally be described as "play," and they expend a great deal of energy doing it. But what are they really doing—and why?

IS PLAY AN ADAPTATION?

The fundamental problem for ethologists studying play behavior is, as we've said, that it doesn't appear to have an obvious function. If that is right, it poses a profound challenge to the fundamental ethological premise that behaviors are products of natural selection. Remember that the logic of the Darwinian story of evolution is that selection favors individuals who move and act in a particular way because the functional effect of the behavior is to confer a selective advantage: it enables the animal to live long enough to produce successful offspring. We expect to be able to observe (or infer) and measure some immediate benefit: a foraging activity leads to the acquiring of calories that provide energy to drive the machine; a hazard-avoidance motor pattern reduces an imminent threat or risk to life; a reproductive act culminates in the successful fertilization of an egg. When you look at playing dogs, you do often see behaviors that resemble (parts of) the adaptive motor patterns that are associated with these functional activities. Chasing and biting, for instance, are commonly seen. But in play the functional goal of the motor pattern isn't attained: a dog that chews up a slipper gains no caloric benefit from doing it.

So what is the benefit of play? Why would any young animal expend a considerable—sometimes an extraordinary—amount of energy in playing if there is no adaptive payback in life? Could play behaviors have arisen for reasons other than as adaptive products of natural selection?

There is a long history of speculation on the subject, from many points of view, among both biologists and psychologists. One approach holds that carefree young animals play simply because it is pleasurable to do so—it feels good to play. Neuroscientist Jaak Panksepp, for example, suggests that play is driven by a special (and evolutionarily very old) "primary" emotional system in the brain that produces a positive affective (emotional) response. We agree that there is good reason to believe that animals derive pleasure from play—indeed, they do from all of their motor activities. Consider reproduction. A copulating pair of dogs may be reproductively successful in the end, but in the moment of sexual interaction the participants don't know or care that their performance will culminate in pregnancy, puppies, and descendants. It is hard enough to teach that lesson to our own overpopulated species. The reward for performing reproductive behaviors is that expressing any kind of intrinsic motor pattern is rewarding in itself—it is self-satisfying.

Most, if not all, motor activity generates brain endorphins—"endogenous opioids" that, like the runner's high, produce a positive internal state. When an animal displays a foraging behavior like EYE > STALK > CHASE, it's reasonable to think that the sequence is performed not because there might (but might not) be a food reward in the end—but because in the moment, there is some sort of pleasurable feedback from simply doing it. A nonhuman animal may or may not have any conscious "felt experience" of it (see the next chapter), but all it has to do to get the reward is to find the right environment and an appropriate releaser. Natural selection may have associated motor patterns with pleasurable side effects (at least in the higher vertebrates) because doing so ultimately contributes to leaving more offspring to the gene pool, even when there is no other immediate reward or goal in sight (or in "mind").

But this kind of answer still begs the thorny question of why play behavior—the actual shape of motor activity that we perceive to be "playful"—would itself have arisen in the course of evolution. Is play, indeed, a specially selected set of motor patterns, of shapes moving in space and time? If that is so, as we have reiterated, we need to identify its selective advantage. Did young animals that play receive a reproductive benefit that was denied to animals that didn't play? Could the satisfaction of "drive pleasure" in and of itself—say, by simply engaging in random movements or interactions with objects—enhance an animal's reproductive fitness? Why, then, is play generally more frequent in pre-reproductive juveniles rather than in sexually mature adults (though it is certainly seen in adults dogs and wolves, too)?

An alternative notion, first advanced by Herbert Spencer in the nineteenth century and more recently termed "surplus resource theory," is that healthy growing young animals have an excess of energy—more available metabolic resources than they need—and need to burn it off through play. Perhaps any kind of random movement might do. This is an interesting argument because growing animals that are fed by their parents never know how much food they are going to get and what the quality in calories or essential nutrients might be. Animals do three things with food: use it for basic metabolism (fuel to run the machine), use it for growth, or store it as fat for future use. Too much fat (and excess weight) is generally maladaptive, however, and too much unburned sugar can send an animal into diabetes-like sugar shock. It's not implausible to think that it is adaptive for an animal with an overly rich food supply to have to "run it off."

There may be something to this idea. Dietland Müller-Schwarze once pointed out that deer fawns that feed on low-quality milk play less than those on high-quality milk. The deer with poorer milk spent more time feeding on grass and therefore didn't have as much time to play. That supports the surplus resource idea. We've had problems with ranchers who overfeed their livestock-guarding dogs. Normally such dogs would sit placidly on guard, not moving much but looking large and threatening to potential predators. The overfed dogs, however, would end up poking around and bothering sheep, like the deer we saw playing with the turkey.

One problem with the resource hypothesis, however, is that any kind of motor activity, like simply running, should be good enough to allow for an animal to balance its energy utilization. It need not "look like" play. In addition, a great deal of play behavior is social, performed with a sibling, another partner, or a group. How likely is it that all the animals in a play interaction would have the same energetic needs at the same time?

Could it simply be that play in and of itself meets an animal's fundamental adaptive needs? Even though play behaviors don't seem to be specialized motor patterns, might they nevertheless somehow help an animal to feed, avoid hazards, or reproduce?

Consider a cat that has just captured a mouse and then goes on to toy with it. We've seen one do that for an entire energetically expensive hour. This sort of behavior certainly bears a strong resemblance to foraging—the cat displays EYE > STALK > CHASE and BITES the mouse. However, the object of this "playful" activity is often left uneaten: the functional end point of the sequence, CONSUME, is absent. There is no food reward here—just, perhaps, the self-satisfaction of expressing some of the motor components of a full sequence. Essentially the same story might be told about human children playing for hours with make-believe baking ovens and plastic toy foods that can't be consumed. A child might go through all of the steps that are necessary for the real activity of baking bread—but no bread gets made and there isn't even a chance that any bread will be made. It looks like the child is playing at an adult activity but without a functional end point or any direct adaptive benefit.

Nor is play likely to assist young animals in avoiding hazards. Any play behavior is energetically expensive and, all things being equal, it will leave an animal with less energy to engage in other essential activities like escaping from a predator. Moreover, play activity between conspecifics can itself cause significant harm. We once observed a litter of four-week-old Border collie pups yelping, rolling over, chasing, and gently nipping one another in what looked like playful abandon—until one of them bit another in the abdomen. The injury was fatal. Similarly, it is true that play behavior sometimes does have a sexual character—think of a young dog enthusiastically rubbing its genitals on your leg—but it doesn't aid in reproduction. That kind of behavior (addressed to the wrong species, at that) can't lead to successful mating and reproduction in the immature juveniles who indulge in it.

If there isn't any immediate fitness benefit to be gained, could it be that the adaptive payoff is deferred? Since it so often resembles aspects of adult behavior, people often suggest that play might constitute a form of "practice" for adulthood. Play might make you a better adult with a greater chance of reproductive success. Perhaps it helps young animals to learn how to hunt better, to deal with aggression more effectively, or to interact more successfully with potential sexual partners.

This is an appealing commonsense idea and an explanation that has been proposed by ethologists and psychologists alike. But some of its implications are rather hard to swallow. Consider hazard avoidance. When a young deer gambols and runs in "play mode," the activity might well strengthen its muscles and stamina, which could enhance its chances of surviving a puma attack. Does this mean we can regard it as practice for effective escape behavior? We don't think so. Like a neonatal alarm call, real escape behavior has to be done just right the first time a fawn encounters a hazard. An incremental improvement in response during development would be of little use—that is precisely why motor patterns are intrinsic, stereotyped, and automatic. When a new mother chews off a newborn's umbilical cord at the proper length the very first time she gives birth, it is a precise behavior that is essential to her pup's survival. No practice is ever required.

Often the argument for practice centers on foraging behavior. Wolves, for instance, don't become effective hunters until they are two to three years old. It's true that as youngsters they follow their parents around, accompanying the hunt at the end of their first year and long into their second. Lots of us who have hunted with village dogs around the world or who train our own dogs know that even with the best breeding it takes at least a couple of years for an animal to become a competent hunter. So experience and learning may well play a role in the improvement of some behavioral patterns—but it doesn't necessarily follow that play behavior in and of itself contributes to proficiency in the adult activity. After all, play only partially resembles adult motor-pattern sequences. The functionally necessary final steps of adult behavioral sequences (e.g., actual consumption or copulation) are rarely expressed in play. It's as if you were practicing to catch a ball by noticing that it was thrown, watching its trajectory through the air, keeping your eye on it as it approaches your hand—and then never actually catching it. Practicing some but not all of a behavior—often doing it in a disorganized sequence and rarely accomplishing a functional result—doesn't seem like a very effective way of learning how to become a better adult.

THE PLAY BOW

So whatever play may be, it does not appear to us that its behavioral character is easily described in classic ethological terms as a set of adaptive motor patterns. That said, ethologists have long observed that when play is initiated, dogs and other related canids exhibit an apparently special and distinct behavioral shape—a "play bow"—which Marc Bekoff, a canid behaviorist at the University of Colorado, characterizes as "a specific and highly stereotyped signal." This is, he says, a highly ritualized action "used almost exclusively in play"—a motor pattern. When a dog intends to play, the story goes, the front of its body and head are lowered close to the ground with front paws extended forward, and the hind quarters are elevated with a lowered tail (as in fig. 38).

The play-bow posture is a species-general characteristic of dogs, and an essentially identical homologous behavior is also found in closely related wild species. This is just what we would expect if the play bow were in fact a distinct intrinsic motor pattern and a product of natural selection—a hazard-avoidance signal, one might say, indicating that what follows will in fact be playful and harmless activity and not a prelude to real aggression or predatory behavior. If evolution indeed gave rise to such special functional signals, it is tempting to conclude that play behavior as a whole might also be an adaptive outcome of natural selection.

The play bow has been interpreted not only as an adaptation but as a cognitively meaningful expression as well—providing evidence that animals might exhibit what philosophers of mind and language call intentional states. These are mental representations that are said to be about things and to reflect "attitudes" like belief and desire. An animal that simply expresses a motor pattern, without having any mental states associated with it, is said to exhibit zero-order intentionality:

FIG. 38 THIS WOLF IS PERFORMING A PLAY BOW. SOME THINK THIS IS AN EVOLVED MOTOR PATTERN SIGNALING THAT ENSUING BEHAVIOR IS INTENDED TO BE PLAYFUL. WE SEE AN ANIMAL IN CONFLICT. PHOTO BY MONTY SLOAN/WOLF PARK.

nothing "in mind." If an animal has first-order intentional states it can represent its own belief or desire—"I *want* to play." With the capacity for higher-order intentionality, humans, for instance, can attribute desires or beliefs to others: "I want *you* to want to play." When your dog drops in front of you, so engagingly and entreatingly, it isn't hard to imagine that this is exactly what it "has in mind."

But we wonder if the so-called play bow in fact really has any adaptive, let alone cognitive, significance. Our student John Glendinning once designed a controlled experiment in which Border collies were confronted with normal and drugged roosters. The dogs typically approached a chicken in EYE > STALK. If the bird ran, the collies went into CHASE—all was normal. But when a rooster failed to run (because it was drugged), the dog got stuck in EYE > STALK even if it had passed the threshold distance that normally would trigger CHASE—much like a cheetah that is unable to chase newborn calves because they don't move. The collie would then exhibit what looked for all the world like a play bow, and dance around the rooster in that position, all the while

FIG. 39 THE DOG IN THE BOTTOM IMAGE IS A SHEEP KILLER. IT HAS GONE THROUGH EYE > STALK, BUT THE SHEEP DIDN'T RUN—SO THE DOG CAN'T CROSS THE THRESHOLD TO CHASE. THE DOG IS IN CONFLICT AND CANNOT PROCEED. LIKEWISE, THE WOLF IN THE TOP IMAGE HAS GONE INTO EYE > STALK, BUT THE BISON DON'T RUN. IN BOTH CASES, WE SEE THE SAME RESULT—IDENTICAL TO THE PLAY BOW. WE ARE NOT INCLINED TO THINK EITHER THE DOG OR THE WOLF WERE INVITING ITS PREY TO PLAY. THE PROBLEM FOR THEM IS THAT THEY JUST CANNOT CONTINUE WITH THE FULL PREDATORY MOTOR-PATTERN SEQUENCE. PHOTOS BY MONTY SLOAN/WOLF PARK (*TOP*) AND JOHN GLENDINNING (*BOTTOM*).

barking. In the same vein, one of John's subjects, an Anatolian shepherd he called Hank, was a known sheep killer who displayed the play bow when his target prey didn't respond appropriately by escaping (fig. 39).

Was Hank thinking: "I want this sheep to believe that I want to play"? Was he "hoping" to deceptively signal to the prey that his subsequent behavior would be harmless? We don't think any such mentalistic assumptions are either warranted or necessary to explain Hank's behavior. (See chap. 10 for a longer discussion of mind and cognition.) A much simpler account, on our view, is that the play bow occurs when an animal is in a temporarily indeterminate state. It can't proceed because it isn't getting an appropriate response (the sheep doesn't run) or because it has directed its behavior at an inappropriate stimulus (an inanimate object or a member of another species, like a turkey, whose signals it cannot properly interpret).

In short, the "playing" animal is in conflict about its next move—and the play bow in fact looks just like a combination of multiple conflicting behavioral shapes. The lowered front end of a play bow is essentially identical to the posture of a canid moving toward prey in EYE > STALK; the raised hind quarters and rear legs are readied for quick flight. Like barking, we think that the shape of the play bow is a result of the animal being in two motivational states at once: it is moving toward a prey object but unable to transition into the normal next step of the predatory sequence.

So we don't believe that a play bow is a special adaptive (let alone intentional) signal at all. We submit that it is an emergent effect of a dog (or wolf) simultaneously displaying two motor-pattern components when it is in multiple or conflicting states. The informational uncertainty of this emergent combinatory event could well attract the attention of a receiver and increase the chances that it would engage in some way with the sender. When it is directed to conspecifics—but almost certainly not to chickens—this could facilitate a social interaction that looks like play. If this is the right way to think about the so-called play bow, however, it shouldn't be interpreted as an adaptive signal generated by natural selection to initiate play. Nor should it encourage us to conclude that play as a whole is also adaptive.

As we suggested at the start of the chapter, we think there is good reason to believe that, like the "play bow" itself, the general character of play behavior is best understood as a result of emergence: a complex product of the interaction of simpler parts. To understand this approach we need to return to the story of development.

GROWTH AND PLAY

Seeing play as a kind of immature adult behavior—in effect treating the juvenile simply as a smaller (if growing) playful adult—is to look at young animals through an extremely narrow lens. Of course they do much more than play. Robert Fagen estimated that more than 90 percent of a typical juvenile mammal's time is spent in nonplay behavior. It's true that some primates may spend more than half of their waking time in play, and humans continue to play for many years, sometimes into old age. An overemphasis on play as a "hallmark" behavior of young animals, however, misses the critical point—that play is never the only thing that a young mammal does, or even most of what it does. The majority of its time budget is dedicated to other activities in the interest of other functional requirements. In order to understand play, we need to look at the bigger behavioral picture—and at the larger story of how a mammal's shape and behavior change over the course of its ontogeny.

In earlier chapters, we made it clear, we hope, that the behavioral landscape changes dramatically over the lifetime of dogs and their canid relatives. Foraging and feeding motor patterns, as we have seen, are substantially different in newborn pups and adults. At each stage in life, mammals are adapted to a distinct niche, both in their shape and their behavior. Embryos in a fertilized egg live in a radically different world from both neonates and sexually mature adults. They all differ in size and shape, and each form of the (same) animal has its own ethogram. Whatever else may be true about their behavior, neonates and adults have distinct sets of intrinsic, stereotyped, and fixed motor patterns. Far from simply being a small, immature, and unformed adult, the neonate is itself a highly specialized organism with a unique shape and its own adaptive feeding and hazard-avoidance patterns, such as nursing, the LOST CALL, and other care-soliciting behaviors. None of these are seen in the adult stage.

It might even be fair to say that the neonate has a more highly evolved shape than the adult does. An adult wolf is a generalized carnivore with nothing much special about its predatory behaviors. Indeed, some insects have foraging motor patterns that look very much like a wolf's. But the mammalian neonate is a more recent creation. It is not only new in geological time, it is also a specialist adapted to a new niche.

We tend to see adults as the ultimate specialized form of an organism—a progressive "improvement" over an unspecialized and helpless newborn. But ontogenetic growth in mammals isn't linear or isometric. That is, the infant shape doesn't simply get larger in all respects. As an animal goes through new stages, it redefines its structure, sometimes even adding on or eliminating an entire organ system, as for instance when the newborn detaches from the placenta at birth. At each stage, the animal's structure—its skeleton, musculature, and nervous system—is adapted to its changing needs and determines a new range of behaviors.

The maturation of a neonate into an adult isn't a simple stepwise affair, the sudden appearance of a new form. An infant's skull isn't just replaced by an adult skull nor does it simply enlarge into an adult's. Rather, the infant's bone is rather dramatically remodeled: it is resorbed and reshaped as it grows, as the images in figures 40 and 41 clearly illustrate.

A mammal's fundamental problem is that it has to make a profound and far-reaching change from being a dependent neonate—whose structures and behaviors are in large part a function of how its parents and other adult caregivers are built and act in response to it—into an independent adult in its own right. The adult has a new form and a new behavioral repertoire: courtship, adult foraging, the ability to escape from predators, territoriality, and parental behavior toward its own offspring. So the neonate's behavior, like its skull, ultimately needs to be remodeled—in effect it has to undergo a metamorphosis into an adult shape. Not unlike the transformation of a caterpillar into a butterfly, the mechanisms of the newborn mammal are partly disassembled and then ultimately reconstructed into the adult shape.

We often call this transitional form a juvenile or an adolescent. These terms suggest a distinct (and perhaps distinctly adaptive) developmen-

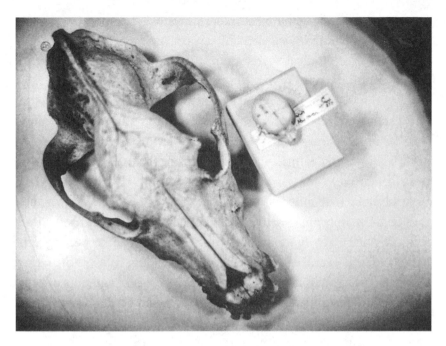

FIG. 40 THE PUPPY SKULL ISN'T JUST SMALLER—IT'S A TOTALLY DIFFERENT SHAPE.
PHOTO BY RICHARD SCHNEIDER.

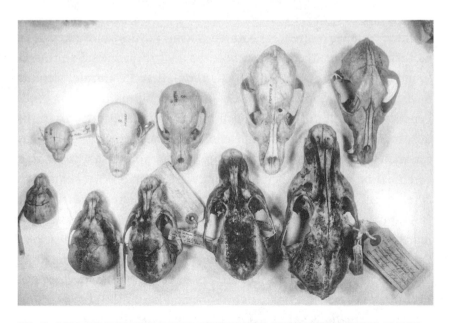

FIG. 41 THE MANY DIFFERENT SHAPES OF DOGS (*TOP*) AND WOLVES (*BOTTOM*) AS THEY
CHANGE FROM NEONATES TO ADULTS. PHOTO BY RICHARD SCHNEIDER.

tal stage. However, mammalian metamorphosis is more than simple development through well-defined stages. As ontogeny unfolds, the animal's entire physical system and attendant behaviors must constantly be reintegrated so that the organism can continue to work as a functioning whole. In effect the adolescent is in the position of someone who has to live in the house that is being remodeled by carpenters who are busily taking it apart and adding on. Juvenile mammals—the archetypal playing animal—are organisms in the midst of this radical transformation.

Remember that intrinsic motor patterns have onsets and offsets: they begin to appear at particular times in an animal's life, and sometimes they cease to function over time. Thus a weaning pup at eight weeks old, say, is still expressing neonatal suckling rules (though they may be on the wane). It is also beginning to express fragments of adult foraging and feeding motor patterns: chewing and swallowing, for instance. In a metamorphosing juvenile, these onsetting adult behaviors begin to occur ever more frequently, but they often appear to be random in their expression. A great example of this can be seen in the work of Gail Richmond and Benjamin Sachs, who studied self-grooming behavior in Norway rats.

These animals—often caricatured as the filthiest of animals—in fact regularly clean their whole bodies as adults. Richmond and Sachs found that the full adult grooming pattern—first the muzzle, then the face, ears, haunch, and tail in succession—actually unfolds in an ontogenetic sequence: the onset of each motor unit is separated from the next by several days. Until all of the parts of the sequence appear, any one individual motor unit of grooming can be expressed by itself or in combination with any other, in any order. The rats don't yet appear to be efficiently and systematically grooming themselves. Rather, these "bits and pieces" of metamorphosing behavior are randomly mixed and matched: together they look very much like play to a human observer. Paul Leyhausen, in his excellent book *Cat Behavior*, saw the same thing in domestic cats. He says that "the instinctive movements of predation are performed independently of one another by the playing cat, in varied combinations with each other and with activities derived from instinctive systems other than predation."

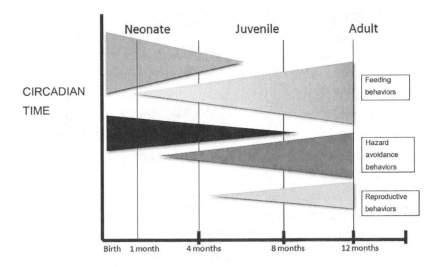

Birth 1 month 4 months 8 months 12 months

ONTOGENETIC TIME

FIG. 42 THE BEHAVIORAL METAMORPHOSIS THAT MAY UNDERLIE PLAY

We think that this picture of play as a random admixture of overlapping behaviors is exactly the right way to understand its fundamental nature—which is to say that play doesn't really have a fundamental nature. It looks much more like an accidental consequence of the expression of other behaviors than a special, unitary adaptation in its own right. Indeed, we see play in mammals as an emergent by-product of their development and life history; in the words of an earlier paper by Coppinger and Smith, "Play is but a term for the protean nature of the neonatal to adult metamorphosis."

The diagram in figure 42 provides a graphical picture of the behavioral metamorphosis that we think underlies mammalian play. The y axis represents circadian time—which means we are looking at typical activity during a twenty-four-hour day. The x axis reflects time through ontogeny—the progress of the life of the animal from birth onward, measured here in months. Each triangle represents a set of intrinsic motor patterns in the three main functional categories of behavior. The thickness of the triangle over time is meant to roughly illustrate how many distinct patterns are present at a particular moment in ontogeny and how often they are expressed.

If you look at a single twenty-four-hour day in the life of a one-month-old pup (the first thin vertical line), you can see that the full complement of neonatal feeding and hazard-avoidance motor patterns are in place. Some components of adult feeding behavior are just beginning to onset; even at three weeks or so a pup can eat some solid foods. By four months (the second vertical line), all of the infant nursing motor patterns have dropped out and any neonatal hazard-avoidance behaviors are expressed at a much lower frequency: the four-month-old is already beginning to exhibit some adult reproductive and hazard-avoidance behaviors as well. At eight months (the third vertical line), all of the neonate's feeding and hazard-avoidance behaviors have offset and some, but not all, of its adult motor patterns have begun to appear at higher frequencies; the dog's first estrus appears as early as seven months. It is not until twelve months (the final vertical line) that a developing dog has an essentially complete set of the foraging, reproductive, and hazard-avoidance behaviors it will have as an adult.

In any one day in the life of a juvenile, we can observe a variable set of (waning) neonatal and (waxing) adult motor patterns. We think that what young mammals like dogs are doing when they play is that they are incorporating bits and pieces of each developing system into a novel motor sequence. So when we look at play we are not seeing a special and unitary set of adaptive motor patterns but rather an emergent behavioral collage—pasted together from random parts of other offsetting neonatal and onsetting adult systems.

Think of a five-month-old dog pup playing with a plastic toy. It takes the object into its mouth, sucks on a soft part, and then chews on (perhaps even "dissects") another part. The pup is displaying bits and pieces of the waning neonatal suckling motor pattern combined with waxing parts of adult feeding by predation or scavenging. Each of those systems is functional when it is engaged at the right time in development—they are adaptive behaviors in the neonate or adult. When a young animal randomly combines fragmentary parts of neonatal and adult patterns, the result is neither functional nor adaptive. But it does have the (emergent) appearance of a novel behavioral picture that we are inclined to call "play."

It follows that the overall story of play in any given species or breed will depend on *which* motor patterns are available for combination in the course of its development. And the more (fragmentary) motor-pattern sequences we see in an animal's repertoire at a given point in development, the more robust and variable its play will seem. As we noted earlier, our student Kathryn Lord has shown that there is considerable variation in general developmental timing between breeds (and between wolves and dogs). To see how such variability might affect play behavior, we conducted an experiment with a pen full of growing puppies of different breeds whose motor-pattern repertoires differ (fig. 5). These included Border collies from Scotland and three different varieties of livestock-guarding dogs—the Italian Maremma, the Macedonian Šarplaninac, and a couple of Transcaucasian Ovcharkas.

The pups in this experiment were exactly the same age and as near as we could get to the same size at about six weeks. We wanted to control for size because we were concerned that it might have an effect on play behavior (think of a four-foot fifty-pound child and a five-foot eighty-pound child trying to play together at racing or wrestling games). When Border collies and livestock dogs are born they are about the same size, typically weighing half a pound each. (It is remarkable how similar wolves, coyotes, jackals, and dogs are in their size and shape at birth. Indeed pups in the genus *Canis* are almost identical in form, even in the ratio of brain volume to total skull area; coat color is the main source of variation.) So, once matched for size and age (stage in ontogenetic development), the trick was to just let them grow up with two Hampshire College students who observed them from the roof of a nearby shed for several months and made ethograms of their behavior every morning.

Working livestock-guarding dogs are not supposed to have any of the predatory motor patterns, and indeed ours did not display any of them throughout the months they spent in the pen. Border collies (and other herding dogs), by contrast, are expected to show the sequence ORIENT > EYE > STALK > CHASE, and all but one dog displayed them all. The results were spectacular. By the time they were four months old, the pups had split into two distinct "play" groups: livestock-guarding dogs in one and Border collies in the other. It looked as if there were two different species in the same pen. Individuals in each group would play with one another but never with individuals from the other group.

The play groups were determined not by personality or individual preference for a playmate but by their differing available behavioral repertoires. The Border collies played games incorporating EYE, STALK, and CHASE; they ran after one other and jumped at falling leaves and grasshoppers in the pen. The guarding dogs were another story entirely. They never chased one another or stalked imaginary prey. Instead, they sat in furry piles playing mouth games, pulling on one another with teeth and tongues, licking faces, nipping and chewing on the body parts of their companions. These features of play in fact arose from the overlap of available (but waning) neonatal nursing behavior and increasingly frequent adult feeding motor patterns. Since there would be no onset of STALK and CHASE (or subsequent predatory motor patterns) in the guarding dogs, their play is apparently less rich than what we see in Border collies. It's not a matter of having a less happy and playful disposition but, rather, a more limited behavioral repertoire. Different styles of play emerge depending on which "simple parts" interact.

When we observe wolves, we see a similar picture. Wolf puppies are often noticeably more robust and varied in their play routines than dogs of the same size and age. This means, according to our hypothesis, that they should have more available motor patterns than the dogs do. That is in fact the case. Juvenile wolf pups, as we have noted, display a food-begging behavior that is perhaps as complex as the EYE > STALK > CHASE > BITE behaviors of the adult. When we watch wolf pups at play, we often see parts of this juvenile foraging pattern in the mix. As a rule, dog pups don't incorporate a FOOD-BEGGING motor pattern into their play—because they don't typically engage in food begging (fig. 43). Usually the mother doesn't have much motivation to regurgitate food (which is elicited in wolves by FOOD-BEGGING) because the pups can feed themselves either from a food dish or "in the wild" at dumps. Therefore, REGURGITATE drops out of the dog mother's repertoire very quickly and FOOD-BEGGING does as well. As a result we don't often observe it in dog play.

So when you see a young dog exuberantly tossing around and chomping on a Frisbee, keep the canid ethogram in mind. Much of this play-

FIG. 43 THESE WOLF PUPS ARE SOLICITING A PARENT FOR REGURGITATED FOOD.
PHOTO BY MONTY SLOAN/WOLF PARK.

ful behavior clearly contains actions that look exactly like EYE, STALK, CHASE, POUNCE, HEAD-TOSS, GRAB-BITE, CHEW, and so on—units of the adult canid motor-pattern sequence whose function is to capture prey. These begin to onset after the first few weeks of life, and that's just when this sort of play behavior first appears. The dog may also MOUTH and LICK the Frisbee: these are still-active components of neonatal suckling. It may try to chew it to pieces as well, a unit of the adult feeding pattern that is taking over. Seen in this light, object play is clearly no unitary behavior—as we've said, it's a pastiche of other onsetting and offsetting behaviors.

In the same vein, when a frisky pup repeatedly approaches a strange novel object, growls, and then jumps back and away from it again, it is expressing parts of infant exploratory behavior and components of adult hazard-avoidance behavior. All in all, play in dogs is a disorganized affair that sometimes does look something like adult functional behavior—but it often contains neonatal behaviors as well. Its elements are out of context, out of order, and generally "out of touch"

FIG. 44 A PERFECTLY GOOD FOREFOOT-STAB, NORMALLY A PART OF A PREDATORY
MOTOR SEQUENCE, DIRECTED AT A BUBBLE UNDER THE ICE. THIS LOOKS LIKE PLAY
BECAUSE IT GETS REPEATED OVER AND OVER-OR DID THE BUBBLE RESEMBLE A MOUSE
ENOUGH TO ELICIT A PREDATORY RESPONSE? PHOTO BY MONTY SLOAN/WOLF PARK.

with the normal adaptive function of the activity. Given that all this
structural and functional oddness, why do people persist in feeling that
play is a recognizable kind of special behavior? It is, we think, an effect
of its emergent character: a sense of the phenomenon that arises from
how we perceive the interplay of its parts—much as the interaction of
hydrogen and oxygen molecules, when they combine into H_2O, causes
us to feel that water is wet.

Another curious property of play in dogs (and perhaps of play in
general) is that it can be so repetitive and persistent—sometimes it
seems almost obsessive-compulsive in its character. Think of children
playing catch, doing little more than tossing a ball back and forth, over
and over again. Remember that a Border collie goes into CHASE when
it reaches a certain threshold point in STALK. In a wild canid, CHASE
will normally stop when the prey is reached and GRAB-BITE will ensue.
But that next step typically isn't expressed in the Border collie—nor
are subsequent units like KILL-BITE and DISSECT—so the functional

end point of the full behavior is never reached. In the adult wild canid's predatory sequence, DISSECT > CONSUME acts as a stop signal. But, like a looping piece of ill-written computer code that can't terminate, a playing Border collie (or a young wolf that hasn't yet developed the full sequence) simply goes back and starts over. Throw a ball in front of a collie: it will chase it and wait for you to throw it again—and again and again. This is nonfunctional and energetically very expensive activity to be sure, but a metamorphosing animal may have no choice but to engage in such apparently maladaptive behavior. The developing motor units have become integrated components of a growing but incomplete biological machine: they are already turned on and operating even though, in the midst of metamorphosis, they don't yet accomplish anything.

Moreover, as we suggested earlier, incomplete or even maladaptive behavior can be a satisfying activity. As we noted, the expression of any motor pattern, quite aside from any adaptive value it may have, provides its own neurohormonal reward. The functionless repetition of an incomplete sequence, even the random performance of overlapping partial motor patterns, may reinforce itself because it offers a constant repetitive stream of self-satisfying activity. No wonder we think of play as "fun".

THE VALUE OF PLAYING

The bottom line is that we can view play as an emergent consequence of the expression of a developing repertoire of motor patterns that are not yet fully functional in the standard adaptationist sense of conferring fitness by directly enhancing an animal's ability to feed, avoid hazards, and reproduce. It is possible, however, that play does confer other important benefits that are indirectly beneficial in an animal's life.

First, consider the growth of a young brain—an essential part of the mechanism that allows for adaptive behaviors. When a dog (or a human child) is born, it has just about as many brain cells as it is ever going to have as an adult. At birth, a pup's neurons are housed in a braincase with a volume of about eight cubic centimeters. By the time it has grown to adulthood a large dog may have a brain volume of around one hundred cubic centimeters (and a larger braincase). The difference in size doesn't result from the production of new neurons; rather, existing cells grow new connections, and new glial cells appear to support bigger networks. How large the pup's brain grows is a function of how many new connections are made—and much of that neural growth depends on the richness of the animal's experience: it is an accommodation to external forces. What is especially exciting about this story of brain growth is that, in dogs, 80 percent of the volume increase occurs after its brief neonatal period and before the dog is four months old. By the time it is eight months old, the dog is a reproductive adult. The biggest brain size increase comes when the dog is an early juvenile— precisely when play emerges.

One intriguing hypothesis, first advanced by John Byers at the University of Idaho, is that play has a critical role in the proliferation of new neural tissue, in stimulating and shaping mammalian brain growth— and having a bigger and more densely connected brain means that an animal will generally be more capable of meeting the adaptive challenges in life. On this view, play does have a crucial (if indirect) value in the early life of an animal—not learning to be an adult or practicing adult behaviors but growing a brain that can better support those behaviors. This hypothesis is supported by considerable neurophysiological evidence recently amassed by Sergio Pellis and others.

There may be a second equally important—but also indirect—benefit of play. Adaptive behaviors unfold one by one over ontogenetic time. Many of them, like canid predation, are highly complex sequences of motor patterns: each part must first develop properly and then be connected together with others, in the right order. If an animal somehow loses early developing motor components that are not yet linked into those sequences, the complete sequence will never appear.

When a motor pattern isn't used, it's often the case that it disappears entirely from the behavioral repertoire. Infant SUCKLING is a good example. If a newborn pup or lamb doesn't find the mother's teat and begin to nurse in the first few hours of life, it will never be able to do it. The unused motor pattern drops out of the system. We saw a related phenomenon in our livestock-guarding dogs. Some of them (atypically) began to exhibit CHASE as very young pups; if it persisted and linked up with EYE/STALK or GRAB-BITE, which have later onsets, the pup turned into an untrustworthy and ineffective guardian. When cooperating farmers and ranchers in our long-term study told us they had problems with CHASE in young pups, we advised them to house the dogs where they could see and smell sheep but could not run. Often CHASE would disappear. When the motor pattern wasn't exercised sufficiently, it dropped out of the repertoire entirely. We think that an important indirect adaptive benefit of play is that it keeps intrinsic motor patterns "alive" and functioning until the entire adult sequence is in place.

Like most other approaches to play, however, these accounts of indirect benefit don't themselves explain the actual behavioral shapes that we see in playing animals. Benefits there may be, but they didn't arise in the course of evolution as a consequence of natural selection pressures on a special class of "play motor patterns."

Is there an evolutionary story to tell at all? Yes. The "evo-devo" perspective of Gould and others reminds us that evolution can operate directly on the timing and character of development itself, not just on specific forms of structure and behavior. As we see it, it was the evolution of the general mammalian pattern of development that gave rise to the possibility of play in dogs. Perhaps the indirect benefits of emergent play contributed reinforcing selective pressures on the evolution of that developmental program. But play behavior in and of itself, on our view, is not an intrinsic behavioral property of dogs or other mammals—not a special evolutionary outcome shaped by direct selection as a way to practice adult behavior, or as a mechanism to provide pleasure, or a means of reinforcing the domestic bond between dogs and humans. Rather, we think, the seemingly mysterious and "protean" nature of play in mammals like the dog is a fortuitous emergent consequence of the development and interaction of other behavioral systems. It arises from the random combination and recombination of fragmentary components of behavior that appear during juvenile metamorphosis—an emergent "by-product" that comes about from the interplay between simple, fragmentary behaviors turning on and off at overlapping times in development.

10 MINDING THE DOG

Dogs and wolves, like many other animals, sometimes store food that they have foraged by digging a "cache" and hiding it away for future consumption. Konrad Lorenz, in his book *The Foundations of Ethology*, describes the behavior this way: "A wolf in the wild carries the remains of a kill to a covert place, digs a hole in the earth there, pushes the piece of plunder in with his nose, shovels the excavated earth back—still using his nose—and then levels the site through shoves with the nose."

To many people, a behavior like caching seems to be purposive and intelligent. (It's certainly adaptive.) It's easy to imagine that a dog or wolf might have "in mind" the functional goal of digging a hole that is able to contain an object; that it intends the food to be hidden from others; that it is consciously aware of the consequences of its activity. In chapter 8 we saw that group hunting in wolves might seem to cry out for a similar cognitive story—but we argued that this phenomenon can be explained as the emergent result of the interaction between simple motor rules. No consciousness or complex cognition is required. What about caching?

Lorenz reminds us that you often see household dogs (and captive wolves) doing something with all the hallmarks of caching behavior: "A young wolf or dog carries a bone to behind the dining room drapes, lays it down there, scrapes violently for a while next to the bone, pushes the bone with his nose to the place where all the scraping was done and then, again with his nose and now squeaking along the surface of the parquetry flooring, shoves the nonexistent earth back into the hole that has not been dug and goes away satisfied." What, if anything, could be in the mind of a dog that digs imaginary holes and shovels fictional dirt?

This example may recall the little story we recounted back in chapter 6 of our Border collie, Flea, who retrieved a tape recorder that was emitting the infant LOST CALL, and then deposited it in her nesting box as if it were a real pup. Flea appeared to execute the RETRIEVE motor pattern without any awareness or consciousness of the difference between her pup and a mechanical device. It was as if she had no special mental representations, let alone feelings, about them — she was simply and automatically expressing an intrinsic motor pattern.

The same thing can be said of caching behavior. Simon Gadbois and his colleagues at Dalhousie University's Canadian Centre for Wolf Research in Halifax, Nova Scotia, have observed and measured canid caching for many years. In a recent paper, they offer substantial evidence that, while caching is by no means a classical single "fixed action pattern" and can vary in the subtle details of its execution, there is an "intrinsic structure and dynamic organization in the flow and expression of the behavior sequence." This intrinsic structure clearly guides the behavior of a dog burying a bone in a nonexistent hole — surely not a conscious plan for hiding food, an appreciation of what holes are for, or a grasp of the outcomes of its action. Gadbois and his colleagues wisely remind us of the work of P. B. Richard (1983), who studied dam building in European beavers. Richard showed, in Gadbois's words, that they "will engage in their behavior when hearing running water (from a recording), even on dry land and without the immediate presence of water."

Is there, then, any role at all for mental states and cognitive traits in ethological explanation? In this book, we've persistently recruited the metaphor of animal-as-machine because we see the basic character of animal behavior as ultimately shaped by its genetically determined physical plan, in the same way that a machine's action is determined and limited by its form. Human-built machines don't seem to have consciousness, awareness of intentional goals, or the ability to plan intelligently. So does it ever make sense for ethologists to talk about the mind of a dog in the same breath that we characterize it as a biological "machine"? In understanding what makes an animal tick, do we need to appeal to more than intrinsic behavioral patterns, accommodations of those traits to development and environment, and the emergent results of interacting patterns? In the cases we've briefly discussed, such as "imaginary" caching behavior in wolves or pup retrieval in Border collies, it wouldn't seem so. But ultimately the answer depends on what you mean by having a mind.

MINDS, BRAINS, AND MACHINES

Descartes may have believed that animals, as mere mechanisms, are necessarily mindless things. But as the twentieth-century British philosopher Bertrand Russell once put it, the mind itself can actually be thought of as "a strange machine which can combine the materials offered to it in the most astonishing ways."

What kind of machines could Russell have had in mind? Certainly not the clocks of the early mechanical age or the steam-powered locomotives of the Industrial Revolution. These devices, driven by simple mechanical forces, certainly did behave—move in time and space (like animals). But at the end of the day no one really believes they were anything more than sometimes complicated but nevertheless mindless artifacts. We may like to anthropomorphize machines like trains ("the little engine that could") but we surely understand that, from a cognitive standpoint, they are an entirely different kind of thing from us humans, indeed from animals in general.

Why is this so? One answer is that a steam locomotive, for instance, has no autonomy of action and no capacity to change its behavior in response to the world around it. Without its human driver, a locomotive can't navigate on its own, doesn't know anything about railroad timetables, can't speed up to meet a schedule, and can't make a decision about which track to follow at a switching point. Machines like the old locomotives may have myriad complex moving parts—but they don't have brains.

The brain, of course, is a central biological machine part in all of the vertebrate animals and in many invertebrates as well. A complex device built out of neural cells and regulated by hormones, the brain provides an organism with centralized control of its other parts and, ultimately, a degree of potentially flexible control over its behavior as a whole. And so our animal-machine metaphor enters the computer age—where countless machines of many kinds now do have brains of a kind, as well as the capability for autonomous action. Digital clocks and watches keep track of time with a microchip brain rather than mechanical parts and movement (and, ironically for our covering metaphor, they don't necessarily tick). Unlike a runaway locomotive, a "fly-by-wire" Boeing 777 can (in principle) land on its own. Smartphones, without being instructed, can detect where you've parked your car, remember the location, and later remind you where you left it.

How much a computer brain might actually be like a biological one is, needless to say, a matter of highly contentious debate: even simple invertebrates like honeybees or cockroaches have brains with considerably more capability than the chip in a digital watch—and the brains of higher vertebrates, like ours or like a dog's, are at the very pinnacle of biological complexity. But the bottom line is that, if you're not a dualist like Descartes who imagined that minds are something outside the physical realm, there must be some kind of physical (or physiological) machinery that embodies a mind. In biological organisms, the brain provides that link.

There are many, many ways to define and think about "mind," and how it might be instantiated in biological or computer brains; the subtleties of this challenging issue are well beyond the scope of this book. Contemporary cognitive science, however, offers one especially illuminating perspective—that a mind, whether in a machine, a human being, or another animal, is fundamentally an information-processing system.

Such a system is able to acquire and represent knowledge about the world or about its own internal states; it can store and retrieve that information; and it can carry out computations with it—manipulations and transformations of information such as comparisons between two representations. Is this an A or a B? Is X larger than Y? If Z is true, it will cause W to occur. In this sense, a mind can be implemented in essentially any type of physical device. Information can be represented in biological structures like neural cells in brains and nervous systems, in silicon chips and electronic switches in modern computers, or even in the brass wheels, cogs, and moving pistons of a clunking steam-driven nineteenth-century machine like Charles Babbage's "analytical engine," a device that could carry out complex computational operations.

While Descartes and other dualists may imagine that mental phenomena could be made of nonphysical stuff (with a quality that nonhuman animals lack), cognitive scientists and realistic biologists generally agree that this isn't the right way to think about the mind. In fact, every cognitive system must ultimately have some physical basis. The particular physical properties don't matter, so long as there is a structure in which information is represented and manipulated. You can have a machine without a mind, but never a mind without a machine.

From this vantage point, even a digital clock can be said to have a mind implemented in its microchip brain. It is capable not only of displaying the time in response to regular electronic impulses but also of representing, storing, and manipulating a richer range of information. For example, it can know that it has to wake you up every week on Tuesday at 8 A.M., and on its own it can check in with the atomic clock at the National Institute of Standards and Technology in Colorado to make sure that it is representing the exact time correctly. This information-processing capability is an integral part of how this sort of machine behaves. Its cognitive capacity is, of course, extremely simple and very limited—and perhaps not too interesting. Nevertheless, a digital watch, or your smartphone, or even a computerized thermostat can be understood to have a kind of mind—undeniably different in capacity from ours, different in the behaviors that it supports, but a mind nonetheless. The same can certainly be said about a dog, though its mind is likely to be at least a bit more cognitively complex and interesting than the mind of your smartphone.

COGNITIVE ARCHITECTURE

The overall form and functional organization of an organism's (or machine's) mind is sometimes referred to as its "cognitive architecture," the system of traits that enable the acquisition, representation, processing, and use of information. The details may well vary from animal to animal, or machine to machine, but the general shape of information-processing systems is universal—and in the case of animals, it no doubt arose quite early in biological evolution. On this view, the mind is part of the animal's phenotype, the conglomerate whole picture of its properties, including its behavior.

At the front end of any information-processing system are input mechanisms that acquire information. In animals, these are the perceptual organs that sense physical events—light and sound waves, pressure changes, and concentrations of molecules—and convert (transduce) them into informational signals that tell the organism what is happening in its environment. These mechanisms for acquiring visual, auditory, olfactory, and tactile information are ubiquitous in the biological world. With few exceptions, even the simplest organisms respond to optical, acoustic, and chemical changes by means of specialized intrinsic adaptive input structures. Many of these are evolutionarily very old indeed. For example, some of the same genes that express pigmented proteins that allow the human retina to detect color differences in light are also found in simple invertebrates from which we diverged a half-billion years ago.

Once it is acquired, sensory information is represented in a central processor—the complex neural structures of a brain. The brain enables processes that allow for an organism to pay attention to a representation long enough to discern its properties, to temporarily hold that information in memory or store it there over the long haul, and to carry out computations with represented information, for example, comparing a stored image of a predator with an incoming representation of an animal that might or might not be a potential threat.

Finally, these central informational states feed into motor systems, providing instructions to the peripheral nervous system to activate muscles and organ systems. These "outputs" of the system give rise to the behaviors that ethologists observe. Whenever we spoke of the releaser of an intrinsic motor pattern, we were in effect characterizing a very simple cognitive state—some representation of visual, auditory, tactile, or other information acquired via sensory input mechanisms—

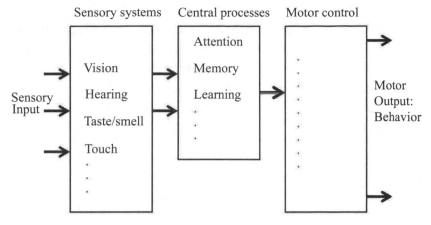

THE 'COGNITIVE ARCHITECTURE'

FIG. 45 COGNITIVE ARCHITECTURE

that triggered action. A simple picture of this "architectural" relationship between information flow and behavior is depicted in figure 45.

Using a term like "architecture" may, however, be a bit misleading. We aren't saying that a mind is itself a physical organ; the schematic in figure 45 isn't the blueprint of a structure that can be precisely located and directly observed in a single part of an organism. The functionality of an animal's mind, its ability to have mental states, arises from the interaction of a complex set of machine parts that represent and use information. For cognitive scientists, "the mind" is the set of informational states that the machine architecture supports. In this sense, it is like a program, the software that runs on a computer—not the machine itself, but the way that the machine represents and manipulates data. A laptop with its operating system erased, and no data stored, doesn't have a mind—nor does a dog without a brain.

The particular neuronal and hormonal properties of the brain as "central processor" are indeed crucial to cognition, but so are the perceptual input organs, as well as feedback wiring to the brain from the entire body via an animal's peripheral nervous system. Moreover, brains are themselves "machine parts" with an extraordinarily complex form: they are by no means single homogeneous organs. They have multiple interacting components, varying cell and tissue types, and highly com-

plex patterns of internal wiring and of neurochemical interaction between wired-together cells. Indeed, like the capacity of an automobile
engine to produce movement, cognition and mind might best be understood as emergent consequences of a host of organismic properties—
together with the informational states they support—acting in concert
to produce something new, complex, and critical to an animal's ability
to behave.

WHY ANIMALS NEED INFORMATION

The heart of the classical ethological view is that behavior is driven by
intrinsic motor patterns—adaptive species-specific rules of action.
And the simple fact is that, for any motor pattern to be expressed, the
animal must have information. It must be cognizant of what is happening in the world around it (and in their internal workings as well). They
may or may not "know that they know" or even "feel" anything about
what they know, but animals, without question, depend on acquiring
and using information to meet their behavioral requirements in life.

If the world were fixed and immutable, nothing ever new under the
sun, biological organisms probably wouldn't need much, if anything,
in the way of information-processing capacity. However, every animal's
environment is constantly changing from moment to moment in a dynamic and often unpredictable state of flux. Food is not always in the
same place; a novel object may be a predator or it may be some other
animal or object that is no threat at all; a potential mate may or may not
be receptive. Under these conditions, effective foraging, hazard avoidance, and reproduction—adaptive responses—can't happen unless the
animal is able to acquire and use information. Is this a food source I can
utilize? Have I actually detected a dangerous predator I should escape
from (a snake, say, and not a moving branch)? Are you a member of my
own species who will accept my sexual advances? It's not at all surprising that sensory input systems (the structures that acquire information
through vision, hearing, olfaction, and so on), and the neurons and
brains that process that information, arose exceedingly early on in the
history of life.

Consider the fear response that dogs and wolves may display to novel objects and events. That response can—and should—be described by ethologists in terms of motor-pattern activity and development (as discussed in chap. 7). From a purely behavioral standpoint, the description of shape and movement, nothing more might need to be said. However, cognition clearly must play a role in explaining the dog's motor response. What, after all, actually makes a stimulus new to the animal? It has to be a matter of detecting, representing, and remembering the informational properties of already known stimuli and comparing them to new ones. It can't simply be an intrinsic response to a fixed sensory input.

Maybe the simple perceptual intensity or the abruptness of a novel event sometimes plays a role in generating a "fear" motor-pattern response. But responding to true novelty—"this is an object, a sound, a person I haven't encountered before"—isn't just a question of directly reacting to a physical stimulus. There has to be some kind of intervening mental (i.e., informational) state, and those cognitive states surely play a critical role in triggering motor patterns. Even the newborn pup's LOST CALL, a great example of an intrinsic, almost reflexive response to temperature differentials, is caused by an informational change and so can be seen as involving a (very simple) mental state. The pup knows that it's cold on one side.

Whether a pup is aware of that knowledge, whether a dog can have a conscious "felt experience" of being lost (or even of being cold) is another matter entirely. Our human brains do certainly seem to give rise to those "sentient" mental states, somehow, and it may be that they are produced in the brains and nervous systems of other animals, which are so similar in mechanical structure and function to our own. We don't rule out the possibility that a dog may have an experience of fear, or feel the pleasure of an endorphin burst when it runs, or sense pain or discomfort when it's injured or stressed. Likewise, it is possible (though we think it's very unlikely) that some nonhuman animals have some kind of conscious sense of "self." But our admittedly conservative assumption is that, however they may actually be experienced by an animal like a dog, many mental states need only be seen as informational "signals" that play a central role in activating the machinery that drives behavior. From that standpoint, it's clear that the "cognitive" in what has recently come to be called cognitive ethology—the study of what animals know about the world around them—is an indispensable dimension of the whole ethological story.

Moreover, like eyes and arms, the fundamental structures of the mind (and the informational capacities they support) aren't acquired through experience: the pioneering ethologists realized that behavior in general must be understood in the light of evolution, and this same conviction is held by the new wave of cognitive ethologists. Whatever else may be said, we expect to find that many (if not all) of the critical properties of animal minds are—like motor patterns—intrinsic traits of an organism that are adaptive consequences of evolution.

A good example is the capacity of many animals to use information about faces. Visual sensory organs are typically in the head, and as animals move in the world, they are likely to encounter others face-to-face. Knowledge about facial characteristics can be highly significant at every stage in life. For instance, in many animals, visual facial information plays a key role in species recognition, allowing the identification of appropriate potential mates, competitors, or predators. Moreover, a great deal of information about an animal's motivational (and perhaps "emotional") state is reflected in facial expression, which often serves as a releasing signal for an appropriate behavioral response. Finally, animal faces tend to be sufficiently distinct from one another that facial differences are able to provide a way (along with auditory and olfactory information in many species) of identifying *particular* individuals, say, members of one's social group or a mother-offspring pair.

Having access to this sort of information can be critically important in newborns, juveniles, and adults alike, and indeed, there is good reason to think that the capacity for acquiring, representing, and processing face information is an intrinsic feature of a species' computational (cognitive) machinery. Quite early in life, human babies pay particular attention to visual stimuli that resemble faces (say, a circle containing spots for eyes, nose, and mouth in the appropriate places), and there is evidence that they engage special neural mechanisms that are dedicated to the task of processing human face-like inputs (as opposed to other objects they encounter visually). Over a lifetime, humans are often remarkably capable of recognizing individual faces even when they have been partially transformed by aging or other factors. This process can be impaired when the neural machinery is "broken"—some individuals, for instance, suffer from prosopagnosia, an inability to recognize faces that arises from faulty development or damage to particular relevant brain structures—but it is otherwise a general species characteristic.

Similar mechanisms for facial information processing have been described in many other species. K. M. Kendrick and colleagues at the Babraham Institute of the University of Cambridge have shown that in sheep, for instance, specialized neural cell populations respond differentially to the faces of sheep of the same breed as opposed to faces of dogs or humans. Moreover, sheep can remember and recognize hundreds of individual sheep faces over long periods of time. Dogs and other canids likewise use information about faces and head shapes: the shape of its mouth (as in the OPEN-MOUTH GAPE), the position of its ears on the head, the direction of the animal's gaze (as in the EYE > STALK motor pattern sequence)—all these facial characteristics have a role in mediating behaviors such as play and social interactions.

There is every reason to believe that this taxonomically widespread cognitive property, like any intrinsic motor pattern, is a product of natural selection. Attending to faces does not have to be learned. It is present in all members of a species, and it has a clear adaptive value in life. Like (other) aspects of physical structure and behavior, such particular cognitive adaptations are built-in features of the overall shape of the biological machine.

ARE DOGS CONSCIOUS?

We now turn to a characteristic of mind that—at least in the human species—certainly has some special and astonishing properties: our powerful sense of "I," our individual self, the fact that we are consciously aware of ourselves and our activity, and the way in which we seem to actually experience the properties of the world (like color or temperature) and of our own internal condition. Being fearful means not only that our adrenal glands produce more norepinephrine, and our heartbeat and respiratory rates increase, but also that these physiological changes *feel like* something to us.

So when humans engage in a behavior like running, our brains can tell our inner I that we are carrying out that activity and not some other. We've said earlier that when a sled dog is racing (or a pup is playing) it probably derives some kind of self-satisfaction from the action itself. Does a sled dog therefore actually feel an exhilarating sense of movement when it runs? Can it be said to "understand" the experience of running? Does a dog apprehend that it is actually itself who is running? These are very deep waters that are especially difficult to navigate when, as ethologists, our first impulse is to directly observe and measure an animal's actual activity—and mental and informational states are famously opaque to that kind of direct observation.

Explaining exactly how a physical brain is able to accomplish any of this, at least in some species, and exactly what consciousness is, remains one of the greatest of scientific challenges. It is a mystery that has remained unsolved since Descartes first posed the mind-body problem many centuries ago. In recent years, cognitive scientists, neuropsychologists, neuroscientists, and biomedical researchers have begun to devise some ingenious ways of trying to get an empirical handle on these matters—but the actual nature of consciousness and its physical basis remain elusive. It's unlikely that it is associated with a single kind of neural shape or structure or that it is just a function of absolute size. Indeed, one reasonable bet is that consciousness is itself a complex emergent result of the interaction of other properties of cognition and brain activity. But whatever the physical basis may be, there seems no doubt that we humans do experience something we call consciousness. And many humans believe—or wish to believe—that animals like dogs do so as well.

It is fortunate that our own species has language—a critical and perhaps unique component of the human central information-processing system that enables us to tell each other, explicitly, what we know, what we're experiencing, and how we feel. Unfortunately, animals like dogs don't offer us this window into the mind. As Bertrand Russell once aptly put it, "No matter how eloquently a dog may bark it cannot tell you that his parents were poor, but honest." Some researchers pin their hopes on the possibility that animal communication systems are more meaningful than we think they are and that decoding the "language" of a dog, for instance, might open the window a bit. The sad truth is that we don't really have much to go on yet if we want to know whether dogs are conscious.

But, in fact, neither conscious awareness nor our ability to talk about it are necessary properties of informationally rich complex behavior. Humans—and no doubt other animals as well—can do some remarkable things in their absence. Think about driving a car. Say you're on your way to work early one morning. As you drive you realize you haven't completed some task and you need to think hard about it before you arrive. We've often had to plan our class lectures this way. Our conscious selves were preoccupied with organizing complex ideas, thinking about questions to ask students, about the assignment for the next class. All the while we were taking in a great deal of information that still allowed us to drive safely and smoothly.

Drivers are undoubtedly in many complex mental states as they perform this very challenging behavior—but in fact you are likely to have little if any immediate awareness of them. We are typically not conscious of the motoric and perceptual events associated with the activity of driving or the quality of the information we are actually utilizing. Miles may go by without your noticing the time passing or the distance traveled. As you drive, you constantly adjust your speed to the flow of traffic, intricately moving the accelerator, brake, and clutch pedals with unconscious precision. Indeed, if you try to think consciously and deliberately about how you're simultaneously manipulating the clutch and gears, for example, you're likely to suddenly jerk the drivetrain even if you've been smoothly driving a standard transmission vehicle for years. Your "I," your human sense of self, may be highly engaged when you're thinking about the day ahead—but not with the behavior of driving itself.

If intelligent humans can engage in such complex behavior without being directly conscious of it, it's not at all implausible to conclude that a sled dog is blissfully unaware of what it is doing when it runs with a team, and a Border collie may go into EYE > STALK without any apprehension of what it is doing or why. We doubt there is an "I" in EYE (if you'll forgive the pun): there need be no "I'm herding sheep" in the mind of a Border collie as it expresses an intrinsic motor pattern. Yes, nonhuman animals have minds-as-information-processors that play a critical adaptive role in driving their behavior. That doesn't necessarily mean they have any explicit consciousness or a felt sense of the experience of their mental/informational states.

ANIMAL GENIUSES?

People frequently tell us that their dogs are geniuses. Perhaps so. But imagine a team of alien scientists hovering over earth in a UFO who have abducted just a single human for study—and he turns out, by chance, to be Stephen Hawking. No matter how carefully they probe him, the aliens would fail in describing our collective level of species intelligence if they conclude that, from a sample size of one, human beings as a species have a good grasp of the mathematics of the space-time continuum (or that humans speak by means of a speech synthesizer).

Casually observing "your dog," or even systematically studying a single Amazon grey parrot that appears to be able to count, or two Border collies that seem to understand a large vocabulary of human words may be suggestive and intriguing—but it simply isn't a good enough basis on which to draw reliable conclusions about a species. One newsworthy "Lassie" may have found its way back after weeks of separation from an owner on a vacation trip who inadvertently left the dog at a rest stop hundreds of miles from home. But no one writes news stories about the 99.9 percent of animals that don't. They usually end up as poignant photos on countless street posters: "Have you seen our lost puppy? Our child is heartbroken."

Dog lovers love to draw rich conclusions about the minds of their animals. The notion that their pets might be "conscious agents" with felt mental states, intentionality, and a sense of self seems to be especially appealing: countless dog owners are delighted to believe that "Rover is happy/sad/loves me . . ." People constantly use the language of complex humanlike cognition in talking about dogs, who are said not just to "know" (in the basic informational sense), but to "know that they know," to believe things, to want and to hope, and even to empathize—to recognize and understand mental states in the humans they live with so closely: "My dog knows when I'm feeling down, and licks my face because he wants to comfort me."

Much of this picture of a dog's mind, we think, is an illusion generated by our deep-seated and persistent impulse to anthropomorphize—by the fact that many of us want to believe that our close animal companions are like us and able to understand us. That illusion can also arise if we don't observe and analyze an animal's behavior closely enough.

A famous example of this problem is the story of Clever Hans, a horse that in late nineteenth-century Germany became well-known for his putative astonishing mental abilities. To all appearances, Kluge Hans, to use his German name, could do arithmetic calculations: add and subtract, solve fractions, and do other tasks that are difficult even for some humans. He provided numerical answers, which also seemed to require the horse to understand linguistic terms like "add" or "divide," by tapping his hoof a certain number of times in response to a questioner. His owner, Wilhelm von Osten, a mathematics teacher and horse trainer, would stand before Hans and ask, "What is three times two?" The horse would tap six times. He also appeared to be able to answer complex questions, for instance, on what day of the week a particular date might fall. Von Osten exhibited Hans widely, acquiring considerable fame (if not notoriety).

Hans's ostensible mental abilities were assessed by a panel of academicians and animal experts who concluded that there was no deception involved in his performance. Von Osten was not a stage magician intent on creating an illusion of intelligence by surreptitiously manipulating Hans's behavior. Ultimately, however, the phenomenon attracted the attention of psychologist Oskar Pfungst, who made several critical discoveries. First, Hans could only appear to solve math problems and answer questions when he was able to observe the questioner. Second, he could only do so when the people who were asking the horse questions already knew the answers themselves. Very subtly, no doubt unconsciously, they were cuing the horse when he got to the right answer. Indeed, if the questioner didn't know the answer to a particular question, neither did the horse. Pfungst concluded that Hans was attending closely to the behavior of the questioner, who gave away the right answer—triggering Hans to stop tapping his hoof at the right number—through small facial characteristics and/or postural movements.

This "Clever Hans effect"—unintentional cuing by the investigator or the experimental environment—remains a methodological confound that bedevils many studies of animal cognition. Cognitive research (indeed behavioral research in general) often requires close contact between animals and human researchers, and it can be exceedingly difficult for investigators to make sure that they haven't somehow inadvertently shaped the behavior of their subjects.

When a study falls prey to these cuing effects, it may still tell us something very important about the information-processing capabilities of nonhuman animals. Learning to respond as Hans did is itself no mean feat for a horse, and it is certainly interesting to find that some—perhaps many—animals are capable of utilizing information provided by humans or other species (we'll have a bit more to say about this shortly). Nonetheless, whenever you hear about a study where an animal gives the impression that it is thinking like a human, you should be very cautious. Like the Wizard of Oz, there might be a "man behind the screen" inadvertently creating a misleading impression of more cognitive sophistication than the animal really has.

All these cautions aside, it is certainly possible to apply appropriate scientific methods to the study of cognition. The last few decades have seen a veritable explosion of serious experimental work on a myriad of questions about animal minds—indeed, dogs in particular have recently become a favorite "hot" research subject. Do dogs experience guilt when they transgress a human's expectations? Do they have a sense of fair play or morality? Do dogs have "intentional states" like believing or hoping? Do they have special abilities to communicate and empathize with their human owners? Does dog barking encode information about the world around the animal?

A host of investigators at newly established canine behavior research centers in places as far-flung as Eötvös Loránd University in Budapest (Hungary) and Duke University in North Carolina are actively pursuing these questions in their labs. Hundreds of papers by researchers in these and other groups have recently appeared in the scientific literature. Their work is extremely interesting, but there are far too many for us in review in detail here—you'll find references to some of them in the bibliography. We'll focus instead on just a few fairly recent studies that exemplify some of the questions that arise and the challenges that scientists face as they try to probe the dog's mind.

WHAT'S THE OBJECT?

Suppose you are sitting in your chair and across the room you see a vase begin to dance on the mantelpiece. You are likely to wonder what makes it do that—because you know that objects like vases can't dance by themselves. Could the movement be caused by an earthquake, or by someone shaking the vase? Or to give another example, when you watch a film and a ghost passes through a wall, you need to suspend your disbelief because our normal understanding is that walls are solid and impassable. People have a rich cognitive appreciation of what particular objects do and what they don't do, as well as a general understanding of the nature of objects.

The "object-ness" of things—the fact that they are not simply individual bits of perceptual information (color, size, shape, sound, smell), but are representations in which those bits are bound together and understood as a unified whole—undoubtedly plays a crucial adaptive role in how many animals, including dogs, think and behave. Prey, predators, and mates all must be perceived and represented as discrete unitary things in the world if an animal is to respond to them effectively and derive a benefit from acting toward them appropriately. So it's plausible to think that the representation of objects is an evolved cognitive trait that is shared by many species.

One particular property of object understanding is robustly evident in humans: once an object is perceived and represented, it is understood to continue to exist even when it can no longer be observed or detected by sensory input systems. The psychologist Jean Piaget called this "object permanence." Even very young human children, by the time they are a year or so old, begin to have this kind of stable representations of things: when we see something in the world we know that it is there and will continue to be there. All normal humans appear to understand things in this way: it's a species-general cognitive characteristic.

Do other animals understand the unitary and permanent nature of objects in the same way? We might well expect it to be true, for instance, of a predator chasing a prey animal that is visible one moment but hidden the next (as it runs in and out of brush while trying to escape). What about dogs? Suppose you show a ball to your dog and then make the motion of throwing it. But you don't let the ball go and instead you hide it behind your back. The dog will likely run out in the direction you would have thrown the ball if you had actually let it go and will search around to see where it went. It certainly seems as if it is looking for the ball and can't find it—as if the dog had a representation of an object that it expected to move through space and appear, as that object, elsewhere. Do dogs understand, the very first time anyone throws one, that an object like a ball won't just disappear? (Did you have to learn that vases don't dance on a mantelpiece of their own volition?) Is object permanence a species-general evolved cognitive trait, an intrinsic property of the information-processing system in dogs?

Not surprisingly, the answers aren't simple. In our classes, we often illustrate the problem with a little experiment. First, an object like a ball is placed in one of three buckets while a dog is watching. Then you put your hand back in the bucket and take the ball out, keeping it in your hand and visible to the dog. Repeat the action with a second bucket, except that this time, when your hand comes out, there is no ball visible in your hand—you've left it in the bucket. Finally, you go with an empty hand to the third bucket, put your empty hand in the bucket and bring it out empty. Now send the dog in and see if he goes to the correct bucket to find the ball. Lots of people have told us that they think that their dogs would easily pass this kind of three-bucket test of object displacement. But out of hundreds of dogs that we and our students have tested, not one ever went directly to the correct bucket and retrieved the ball.

Human children, by contrast, are quite successful at this sort of task after the age of two, and the ability has also been demonstrated in other great apes: gorillas, orangutans, and chimpanzees. In 1992, Sylvain Gagnon and François Doré carried out a version of the experiments that we did; they concluded that dogs have object permanence as well—perhaps surprisingly, since other studies have failed to show it in many animals (other than the higher primates), including monkeys, dolphins, and domestic cats. On closer inspection, however, an Australian team (Emma Collier-Baker and her colleagues) found that Gagnon and Doré hadn't controlled for the Clever Hans effect. Among other things, Collier-Baker showed that the position of the experimenters who were handling the dogs provided an unintentional cue to the location of the displaced object.

Collier-Baker, Joanne M. Davis, and Thomas Suddendorf went on to repeat the study with better controls for unintentional cueing, and still found that dogs seemed to be able to reliably find a hidden displaced object. They noticed, however, that there might be an additional confounding factor—whether or not the hidden object had been put into an immediately adjacent container. In subsequent experiments they found that dogs were only successful at the task when the target object was to be found in a container right next to the one where it had initially been placed. Dogs actually seemed, in effect, to be following a simple rule, "go to the next box over." This was a factor that we hadn't considered when we did our own informal tests and found no evidence for object permanence; we may well have always hidden the ball in a nonadjacent bucket.

It's worth thinking for a moment about why a dog might display a tendency for seeking objects in adjacent places. When animals are collecting information about their environment—as they forage for food, for instance—they might do it by haphazard inspection, by randomly moving about as they explore their environment. But it is likely far more efficient (and energetically less costly) to search systematically, for example, to look successively at immediately adjacent places. Natural selection may well have favored an information-gathering mechanism of this sort, giving rise to a simple intrinsic cognitive adaptation that supports general behavioral requirements.

We certainly don't want to dismiss out of hand the possibility that dogs or other animals might also have more complex mental capacities. Many authors have concluded that dogs do have object permanence, at least in some form. However, subtle but critical issues in experimental design of the sort that we noted above demonstrate how challenging it can be to accurately study these properties in nonhuman animals, and how easily their cognitive abilities might be overinterpreted. There are many such complicating factors. Consider breed differences, for instance. Could these have an effect on the outcome (and interpretation) of cognitive studies? All too often, experimenters recruit a ragtag sample of diverse breeds as their subjects (often the pets of the researchers and their graduate students). However, as we've noted several times from a behavioral standpoint, breed differences do matter, and dogs of different breeds in fact pay attention and respond to objects, especially moving ones, in quite different ways.

Some dogs chase things and some don't. You'll recall that livestock-guarding dogs won't go after a ball that is tossed in front of it—simply because they don't have the CHASE motor pattern. We assume that their visual system, common to all dogs, is certainly capable of detecting a ball. So when one rolls past a Maremma and the dog simply watches it go by, we wonder if the animal has established and maintained any stable representation of the object at all. As far as the Maremma's behavior is concerned, the ball might just as well have disappeared.

In contrast, other breeds such as Border collies will chase balls incessantly. Throw one and the dog will EYE its trajectory and then go into CHASE, and it is likely to be successful in locating the object even when it's hidden from view. So it's quite reasonable to conclude that Border collies (and other herding dogs) do have an object representation of the ball and perhaps the ability to "understand" object permanence. Interestingly, however, we've raised Border collies in isolated kennels with very limited human or other social interaction. In this restricted environment, these animals never exhibited CHASE—and they never paid attention to or chased balls. In assessing studies of the mental abilities of "the dog" we think it's crucial not to minimize the kind of intrinsic differences among breeds that we've found in motor-pattern repertoires or to overlook the accommodative effects of individual developmental histories.

WHAT'S THE POINT?

Consider, as well, what happens when we aim a pointing finger at an object. A human observer will understand that there may be something of interest at the end of an imaginary line from the tip of a human finger: food, perhaps. We acquire a great deal of information in this way. Indeed, "referring" to objects by pointing may have long preceded our use of language. Can dogs understand and use this kind of human-provided information?

Brian Hare and his research group (then at the Max Planck Institute for Evolutionary Anthropology in Leipzig, Germany) suggested that this ability is a special, perhaps unique, intrinsic cognitive capacity that arose because of the close association of dogs and humans and represents a convergence of mental traits and behaviors that coevolved in the context of animal domestication. They collected experimental data suggesting that dogs are quite generally capable of using visual information provided by humans, but wolves aren't. Moreover, dogs seem to be even better than chimpanzees at interpreting the meaning of human pointing cues—in spite of the fact that chimps are often presumed to have a significant measure of something akin to human intelligence.

This claim certainly appeals to the popular notion of a special bond between us and our dogs. It caters to our desire to see intelligence in them, and it has gained a lot of attention in the media. The result is routinely cited in the scientific literature as evidence for a special cognitive ability in dogs. But recent work by other researchers (less often cited) points in a different direction. Monique Udell, Nicole Corey, and Clive Wynne at the University of Florida tested a group of wolves that were intensively hand-raised from birth and showed that these "wild" animals were just as good as dogs—and, in some trials, better—at interpreting a human point. So given the proper environment and opportunity, wolves can rise to the occasion, too. The ability to use human pointing information, therefore, would not appear to be a unique cognitive adaptation that arose in the recent evolution of domestic dogs and does not provide evidence for a special mental affinity with "man's best friend."

Moreover, all of Hare's adult dog subjects were pet animals that had been raised in close contact with humans from an early age. This environmental factor alone could have made all the difference in his experiments. In contrast, Udell, Corey, and Wynne included nonpet dogs in their study, animals that ended up in shelters (like some five million of the seventy-five million dogs in the United States). It turned out that these shelter dogs—animals that were raised from puppyhood in shelters or were abandoned by owners because they were inadequately socialized—were incapable of using pointing information. So even within the species, cognitive ability once more appears to be subject to an animal's life history and its accommodation to experience in the course of development.

Comparing cognition (or "intelligence") between different species like dogs and chimpanzees (or dogs and humans) can be an intriguing exercise. In principle, it can provide us with considerable insight into the nature of the mind and its evolution. In practice, however, it is exceptionally difficult to control cross-species experiments—to minimize any differences between the groups being studied in order to isolate the property that is under experimental investigation.

In Hare's study, for instance, dogs had to find food from pointing cues that were provided in the context of what was effectively their natural environment—living in very close and often continuous association with humans. Moreover, all of his subjects had grown up in somebody's house being fed by humans with their hands. The chimps, by contrast, were captive wild animals. The experimenters required them to reach through a hole in a glass cage enclosure—a far cry from the natural conditions in which their species normally forages for food.

So we're skeptical of the conclusion that domestic dogs have a unique cognitive ability to use human-provided pointing cues or have a special adaptation that was selected for in the context of domestication. There is certainly a growing and very rich body of research on this ability in dogs, but no one should have been surprised to find that they can utilize human-provided information. The notion that dogs will respond to pointing, voice commands, whistles, or gestures is hardly news. Humans have been using this canine capacity effectively for thousands of years. Every hunter, every sled-dog racer, every shepherd, every dog trainer uses hand signals that are no different than a point. Fifty years of research with chimpanzees who have learned manual-gestural signing systems (whether or not they are truly like human language—a highly contentious question) shows that these animals, too, under the right training conditions, can use human-provided information in some remarkable ways. For that matter, a wide variety of other nonhuman organisms, including animals that most people are inclined to regard as far "less intelligent" than a chimp or even a dog—human-raised bats, for instance—also seem to have the ability to use pointing cues. We direct your attention to the images in plate 8, which show Ray directing turkeys with a point, and a human shepherd similarly directing his sheep.

Giving one's attention to other organisms—following their movement and acquiring information from them (as well as from the environment)—is no doubt essential in supporting many kinds of behavior, from group foraging to mate recognition and predator avoidance, in many—if not all—animals. We suspect that this general capacity was one of the earliest cognitive adaptations. Indeed, the basic ability to pay attention to other organisms might well have played a key role in the evolution of cognition. Detecting objects (and ultimately understanding anything about them), for instance, requires an animal to fix its attention on certain properties of the world that are in flux and to pay attention long enough to know that those properties are bound together as a particular kind of thing.

Might dogs nevertheless have special attentional capabilities? A study by Alexandra Horowitz at Barnard College (Columbia University) suggests that dogs pay attention to a wide range of signals from others during play, for example, the "play bow" we discussed earlier, as well as a host of other movements, facial expressions, and postural cues.

Horowitz argues that the way these signals are deployed during play is evidence that dogs are not only paying attention to others, they are also paying attention to whether others are paying attention to them. If a particular movement or signal doesn't cause engagement with the sender, dogs flexibly switch to a different attention-getting activity. Horowitz suggests that a similar picture of "attention to attention" is seen when dogs interact with humans. In much the same vein, Ádám Miklósi and his Hungarian colleagues have reported that dogs are able to attract the attention of humans in order to obtain objects that are inaccessible to them.

How widespread this richer attentional capacity might be—whether it is also found in unsocialized or free-living dogs, in wolves, or indeed in other animals—remains an open question. Udell and her colleagues, for instance, recently found, in their words, that "only pet dogs recognize that a human reading (with a book covering her face) is inattentive; in our study, dogs living in a shelter and hand-reared wolves which had less opportunity to experience this scenario in their current environment, begged equally from a person who was reading and a person looking at them."

Much more work needs to be done, but studies like these do raise interesting possibilities about the nature (and limits) of animal minds. The phenomenon of attention to attention, for instance, is interestingly suggestive of what cognitive scientists have called "theory of mind"—the capacity for apprehending the mental state of others. Humans clearly have this sort of "mind-reading" ability, which might well be a central property of consciousness (or at least a necessary precursor to it). Researchers have long looked for indications of theory of mind in other nonhuman animals, but convincing evidence has been very hard to come by. Perhaps some dogs do have a rudimentary version. Attention to attention in dogs may point to at least a glimmer of canine consciousness.

That all said, dog lovers and researchers alike should be cautious about jumping to hasty conclusions about how much of behavior is actually associated with and guided by complex mental states. Yes, dogs are cognitively capable animals, whose movement in time and space is supported by information about the world. But the impulse to see "man's best friend" as a special kind of animal has created a cultural perception of dogs—charming and appealing as it may be—that can easily encourage an overblown picture of their mentality. Cognitive ethology is beginning to offer important new insights into the minds of dogs and other animals, but we think the jury is still out on just how sentient or how "smart" your dog might really be.

A LAST WORD

The title of this book, *How Dogs Work*, may seem to hold out the promise of a definitive answer to the question of why animals like dogs do what they do. But there is no single simple answer—in large part because dogs don't all work the same way. What is especially fascinating about them as study animals for ethologists is that they come in so many shapes and sizes, and each variant, each "breed," works differently.

When Border collies are used to herd sheep, for instance, they display a very specific posture that is sometimes called "clapping" by working shepherds and herding trial enthusiasts. We characterized the behavior as an intrinsic motor pattern we call EYE-STALK. It's a feature of the inborn shape of a Border collie. You can't teach just any dog how to clap—but you can train a Border collie that claps to herd sheep. Clapping is genetic. Shepherds and breeders pay close attention to the specific shape of the clap for each dog (there is some small variation even in intrinsic features of an animal) and they selectively breed those dogs that have the best clap.

But livestock-guarding dogs like the Italian Maremma rarely if ever display this behavior, and you can't teach one to clap if it doesn't spontaneously display the motor pattern early in life. If an individual livestock-guarding dog does show clapping behavior, shepherds won't use the dog, and certainly never would breed to it—Darwin's artificial selection in action.

If you've ever watched Border collies closely, no matter how much you may love them as a pet or appreciate their ability to help humans work, we think it's hard not to agree with Descartes's centuries-old notion that their behavior is mechanical. Like pistons moving in an engine, the shape of clapping is determined and limited by the way a Border collie is built and how its parts work together. This is, we've argued, a good metaphorical way of beginning to understand the fundamental insight of ethology: that a behavior is a physical trait just like any other taxonomic character of a species (or a breed), a form of the animal that was shaped by biological evolution and arose because that shape conferred a selective advantage in a specific environment.

Do these intrinsic, machinelike traits (which can differ between breeds) explain all of dog behavior? No, biological machines are unlike human-built devices in that their shape—and their behavior—changes over the lifetime of an animal. Border collies, for example, are only at their "trial best" quality between three and six years old. Breeders may have sorted through a population of dogs and allowed only those with the best prospects of becoming a successful herder or a trial competitor to reproduce. But they do not perform perfectly "right out of the box" (as we hope a new computer will). A successful Border collie will need to grow and work in the world around it. Different pressures and different experience will build a different dog—and only certain developing shapes will work well. It probably is the same with natural selection: whatever their initial intrinsic capabilities may be, very few animals ever make it to reproductive age—only 10 percent of wolf pups live past three years of age.

So the story of behavior isn't just a stark dichotomy between an animal's genetic properties or its experience, between nature or nurture. Ethologists have come to realize that behavior is always a synergistic result of both—because it is a function of physical shape, and biological shape is never completely predetermined by genes. Yes, all animals are products of their genes, and sometimes the genes conspire to produce an intrinsic trait that effectively never changes. But many characteristics can be modified, if slightly, as an organism accommodates to its environment. The final shape of a dog, including its behavior, is always a result of how its intrinsic form responds to the properties of the environment in which it grows and lives. Intrinsic and accommodative forces act together in determining how a dog works.

Is that, then, the whole story? For ethologists a basic working assumption has always been that the activity of animals is shaped by adaptive forces of natural selection—behavior evolves. There is, however, a problem. Some very interesting behaviors don't seem to easily fit with that "selectionist" picture. The complexity and subtlety of collective hunting in wolves, for instance—adaptive though it may be—isn't readily explained by appealing to selection pressures for specific movements. Researchers who have tried to explain this remarkable behavior have reached for alternative hypotheses—for instance, that wolves use intelligent problem solving and communication as they interact with one another on the hunt: "cut it off at the pass and I'll ambush it when it comes this way."

But others working on these and similar examples of behavioral complexity—swarming in fish or the flocking behavior of birds—have offered a rather different (and ingenious) sort of explanation: that novel and complex phenomena can "emerge" from the interaction of simpler properties. Mathematicians, computer scientists, and biologists alike have begun to see that the idea of emergence is important in illuminating many aspects of form. We think that it offers an exciting new way to understand some familiar but perplexing behavioral phenomena in dogs—for example, play and barking—that otherwise resist explanation in the standard adaptationist and selectionist framework. It is an important third dimension of the ethological story.

Finally, the story of behavior is incomplete without also accounting for the fact that animals need to know what is going on in the world around them. In that sense, at least, they must have minds: special properties of their shape that can represent and use information to guide action. Animals may be like machines, but they are information-processing machines. Do they also have more complex mental attributes, like sentience and self-awareness? Perhaps. We don't say that your loving pet is nothing but a cold and unfeeling machine. Humans are biological machines too, after all, and we have minds that are conscious of feeling and experiencing the world. Computers and robots are becoming more and more capable by the day, and it's not out of the question that sentient machines will become a reality in the not-too-distant future. We've tried to show, however, that much of the story of how dogs work can still be told without necessarily assuming they have higher-order cognitive properties like consciousness.

You may not agree with our metaphor of animals as machines whose behavior is largely driven by intrinsic properties, or our claim that wolf hunting behavior, dog barking, and play are emergent phenomena that didn't arise as products of evolution through natural selection, or our view that a lot of a dog's behavior can be explained without appealing to consciousness or sentience. Perhaps you find some of these ideas unsettling or contrary to your own experience of dogs. Good! As lifelong college professors, we think that the best way to teach and learn is to encourage active and critical inquiry. We want our students and readers to think like scientists, and we don't want you to take any of our conclusions to be the final word about how dogs work.

In every domain of scientific inquiry you will find multiple perspectives and competing theories; ethology is no exception. But all things being equal, science generally prefers the simplest explanations that cover observable facts, even if more complicated ones might appeal to us intellectually or sentimentally. When you watch a Border collie go through its paces, for instance—clapping and chasing, and directing sheep in what looks like a beautiful and intelligent dance—it's easy to entertain the hypothesis that the dog is aware of its task, that it has conscious goals in mind, that it wants to please the shepherd. As engaging as this picture might be (and many people like to think that the Border collie is the most intelligent of dogs), a far simpler explanation rests on an understanding of the Border collie's intrinsic behavioral mechanisms: the apparent intelligence of its activity resides in the shepherd, not the dog. Without the shepherd "turning switches in the machine"—controlling the expression of Border collie motor patterns—the dog may chase after sheep but it will never display its famous herding behavior. No shepherd ever just turned to a dog and said: "Go over and bring those sheep to the barn. Then meet me at the bar for a beer and a treat."

Understanding the astonishingly intricate interplay between intrinsic genetic mechanisms, accommodations in development, emergent phenomena, and mind remains one of the greatest challenges of science. Whatever may ultimately turn out to be the best story about how animals work, we hope you've come to appreciate, as we do, that dogs—however else people may regard and value them—are wonderfully interesting animals for ethologists to study.

ACKNOWLEDGMENTS

Over the course of forty-five years, much of our work was carried out in collaboration with students and colleagues at Hampshire College (Amherst, MA) and at Wolf Park (Battle Ground, IN). It would be impossible to list all the people who have made significant contributions to our studies. The main thrust came from an idea of Charles (Chuck) Longsworth, an early president of the college, about how to fund an enterprise as big as our livestock-guarding dog project. Sue Mellon of the Mellon Foundation and Bill Dietel of the Rockefeller Brothers Fund were especially generous in the early years. The third president of Hampshire College, Adele Simmons, kept us going with an imaginative fundraising initiative. She introduced us to people like Helenty Homans; together they helped us build a fabulous research laboratory with classroom, office, and kennel space. The kennel attracted both faculty and students to focus on questions about animal behavior from diverse disciplinary perspectives. These interactions generated some fascinating scientific work. Matthew Berry, Susan Goldhor, Ramon Escobedo Martinez, Lynn Miller, Donatella Muirhead, Cristina Muro, Francisco Petrucci-Fonseca, Silvia Ribeiro, Will Ryan, C. K. Smith, Lee Spector, Daniel Stewart, Mike Sutherland, and Dean Arthur Westing, among many others, made invaluable contributions to the research.

The most significant thing for us was the involvement of our students. Hampshire's unique program, emphasizing the importance of self-directed inquiry and interaction across disciplines, stimulated bright undergraduates to do original research and publish their work, and a dozen or more of them have helped us with this book. Many have become professional colleagues and remain our good friends. Their contributions are often cited in the text and the bibliographies. High on this long list (with many, many names regretfully omitted) are Cindi Arons, Liska Clemence, Abby Grace Drake, John Glendinning, Gail Langeloh, Kathryn Lord, Jay Lorenz, Alessia Ortolani, Mike Sands, David Schimel, Richard Schneider, Ellen Torop, Karyn Vogel, and Emily Groves Yazwinsky. Our former student Malaga Baldi was also our book agent, who stuck with us through thick and thin. Our long association with Wolf Park was equally rewarding. Founded in 1972 by Erich Klinghammer, the park has become a preeminent educational and research facility. Over the years, the staff and supporters of Wolf Park have helped with every aspect of our research. Erich provided us with an invaluable setting in which to test and teach our ideas about canid behavior, and he also connected us with the wider world of ethology. Through Erich we have been able to meet and interact with the famous (and many less well-known but equally important) European ethologists, and he arranged for his staff to help in conducting many of our experiments for us. Hardly a day has passed over the past forty-five years that there hasn't been some interaction with the likes of Dana Drenzek, Pat Goodman, Holly Jaycox, Tom O'Dowd, and Monty Sloan.

Beyond these institutional connections, an enormous host of people, far too many to name them all, have been instrumental in supporting our work. Among them are Italo Costa, Hudson Glimp, Carl Phillips, Peter Pinardi, and Paul Trachtman; Peter Neville from the U.K. Centre of Applied Pet Ethology inspired the writing of an early version of this book. Last, but of course not least, we thank Lorna Coppinger and Carol Gomez Feinstein, who made innumerable and immeasurable contributions, both intellectual and personal.

REFERENCES

The following books, journal articles, and other materials were consulted or referred to in each chapter's text. They may be of interest to readers looking for alternative views on the subjects we discuss, or a broader and deeper (or more technical) approach to them.

CHAPTER 1

Burghardt, G. M. 2007. Critical anthropomorphism, uncritical anthropocentrism, and naïve nominalism. *Comparative Cognition and Behavior Reviews* 2:136–38.

Coppinger, R., and L. Coppinger. 2001. *Dogs: A New Understanding of Canine Origin, Behavior, and Evolution.* Chicago: University of Chicago Press.

Coppinger, R. P. and C. K. Smith. 1983. The domestication of evolution. *Environmental Conservation* 10:283–92.

Coppinger, R., L. Spector, and L. Miller. 2010. What, if anything, is a wolf? In *The World of Wolves: New Perspectives on Ecology, Behaviour, and Management,* ed. M. Musiani, L. Boitani, and P. C. Paquet, 41–67. Calgary: University of Calgary Press.

Darwin, C. 1858. On the tendency of species to form varieties and on the perpetuation of varieties of species by natural selection. *Zoological Journal of the Linnean Society* 3:45–62.

———. 1899. *The Variation of Animals and Plants under Domestication,* Vol. 1. New York: Appleton.

Frank, H., and M. G. Frank. 1982. On the effects of domestication on canine social development and behavior. *Applied Animal Ethology* 8:507–25.

Klinghammer, E., and P. A. Goodman. 1987. Socialization and management of wolves in captivity. *Man and Wolf,* ed. H. Frank, 31–59. Dordrecht: Dr. W. Junk Publishers.

Larson, G., E. K. Karlsson, A. Perri, M. T. Webster, S. Y. W. Ho, J. Peters, et al. 2012. Rethinking dog domestication by integrating genetics, archeology, and biogeography. *Proceedings of the National Academy of Sciences* 109, no. 23: 8878–83.

Lorenz, K. Z. 1950. The comparative method in studying innate behavior patterns. In *Physiological Mechanisms of Animal Behavior,* ed. Society for Experimental Biology,

221–68. Symposia of the Society of Experimental Biology, no. 4. Cambridge: Cambridge University Press.

———. 1982. *The Foundations of Ethology: The Principal Ideas and Discoveries in Animal Behavior*. New York: Simon and Schuster.

McDermott, D. 2001. *Mind and Mechanism*. Cambridge, MA: MIT Press.

Morey, D. F. 2005. Burying the evidence: the social bond between dogs and people. *Journal of Archaeological Science* 33:158–75.

Podberscek, A. L. 2009. Good to pet and eat: the keeping and consuming of dogs and cats in South Korea. *Journal of Social Issues* 65:615–32.

Price, E. O. 1984. Behavioral aspects of animal domestication. *Quarterly Review of Biology* 59:1–32.

———. 1998. Behavioral genetics and the process of animal domestication. In *Genetics and the Behavior of Domestic Animals*, ed. T. Grandin, 31–65. San Diego, CA: Academic Press.

Woolpy, J. H., and B. E. Ginsburg. 1967. Wolf socialization: a study of temperament in a wild social species. *American Zoologist* 7:357–63.

World Health Organization. 2004. *WHO Expert Consultation on Rabies: First Report*. Geneva, Switzerland: World Health Organization.

Wynne, Clive D. L. 2007. What are animals? why anthropomorphism is still not a scientific approach to behavior. *Comparative Cognition and Behavior Reviews* 2:125–35.

CHAPTER 2

Burghardt, G. M. 1973. Instinct and innate behavior: toward an ethological psychology. In *The Study of Behavior: Learning, Motivation, Emotion, and Instinct*, ed. J. A. Nevin and G. S. Reynolds, 322–400. Glenview, IL: Scott Foresman.

Coppinger, L., and R. Coppinger. 1982. Livestock-guarding dogs that wear sheep's clothing. *Smithsonian* 13, no. 1 (April): 64–73.

Coppinger, R., and L. Coppinger. 1995. Interactions between livestock guarding dogs and wolves. In *Ecology and Conservation of Wolves in a Changing World*, ed. L. N. Carbyn, S. H. Fritts, and D. R. Seip, 523–26. Occasional Publication no. 35. Edmonton: Canadian Circumpolar Institute.

———. 1993. Dogs for herding and guarding livestock. In *Livestock Handling and Transport*, ed. T. Grandin, 179–96. Wallingford, UK: CAB International.

Coppinger, R., L. Coppinger, G. Langeloh, L. Gettler, and J., Lorenz. 1988. A decade of use of livestock guarding dogs. *Proceedings of the Vertebrate Pest Conference* 13:209–14.

Coppinger, R., J. Lorenz, J. Glendinning, and P. Pinardi. 1983. Attentiveness of guarding dogs for reducing predation on domestic sheep. *Journal of Range Management* 36:275–79.

Coppinger, R., and R. Schneider. 1995. The evolution of working dog behavior. In *The Domestic Dog: Its Evolution, Behaviour, and Interactions with People*, ed. J. Serpell, 21–47. Cambridge: Cambridge University Press.

Dawkins, R. 1976. Hierarchical organization: a candidate principle for ethology. In *Growing Points in Ethology*, ed. P. P. G. Bateson and R. A. Hinde, 7–54. Cambridge: Cambridge University Press.

Eibl-Eibesfeldt, I. 1970. *Ethology: The Biology of Behavior*. New York: Holt, Rinehart and Winston.

Kruuk, H. 2003. *Niko's Nature: A Life of Niko Tinbergen and the Science of Animal Behavior*. Oxford: Oxford University Press.

Lehner, P. N. 1996. *Handbook of Ethological Methods*. Cambridge: Cambridge University Press.

Lorenz, J., R. Coppinger, and M. Sutherland. 1986. Causes and economic effects of mortality in livestock guarding dogs. *Journal of Range Management* 39:293–95.

Lorenz, K. 1955. *Man Meets Dog*. Boston: Houghton Mifflin Company.

Martin, P., and P. Bateson. 2007. *Measuring Behaviour: An Introductory Guide*. Cambridge: Cambridge University Press.

Nisbett, A. 1976. *Konrad Lorenz*. New York: Harcourt Brace Jovanovich.

Price, E. O. 1999. Behavioral development in animals undergoing domestication. *Applied Animal Behaviour Science* 65:245–71.

Spencer, J. P., M. S. Blumberg, B. McMurray, S. R. Robinson, L. K. Samuelson, and J. B. Tomblin. 2009. Short arms and talking eggs: why we should no longer abide the nativist-empiricist debate. *Child Development Perspectives* 3, no. 2: 79–87.

Thorpe, W. H. 1966. Ritualization in ontogeny. *Philosophical Transactions of the Royal Society of London*, ser. B: *Biological Sciences* 251:311–19.

Tinbergen, N. 1951. *The Study of Instinct*. Oxford: Oxford University Press.

———. 1968. On war and peace in animals and man. *Science* 160:1411–18.

———. 1963. On aims and methods of ethology. *Zeitschrift für Tierpsychologie* 20:410–33.

———. 1972. *The Animal in Its World*, vol. 1, *Field Studies*. London: George Allen and Unwin.

Twitty, V. C. 1966. *Of Scientists and Salamanders*. San Francisco: W. H. Freeman.

CHAPTER 3

Arons, C., and W. Shoemaker. 1992. The distribution of catecholamines and β-endorphin in the brains of three behaviorally distinct breeds of dogs and their F_1 hybrids. *Brain Research* 594, no. 1: 31–39.

Coppinger, L. 1977. *The World of Sled Dogs*. New York: Howell Book House.

Coppinger, R. 2009. Physical and behavioral conformation of dogs. *Skripte Animal Learn Internationales Hundesymposium* 2009:5–18.

Haldane, J. B. S. 1926. On being the right size. *Harper's Magazine* (March), 424–27.

Kemper, K. E., P. M. Visscher, and M. E. Goddard. 2012. Genetic architecture of body size in mammals. *Genome Biology* 13:244.

Morey, D. F. 1992. Size, shape, and development in the evolution of the domestic dog. *Journal of Archaeological Science* 19:181–204.

Phillips, C. J., R. P. Coppinger, and D. S. Schimel. 1981. Hyperthermia in running sled dogs. *Journal of Applied Physiology* 51:135–42.

Sands, M. W., R. P. Coppinger, and C. J. Phillips., 1977. Comparisons of thermal sweating and histology of sweat glands of selected canids. *Journal of Mammalogy* 58: 74–78.

Stockard, C. R. 1941. *The Genetic and Endocrinic Basis for Differences in Form and Behavior.* American Anatomical Memoirs, no. 19. Philadelphia: Wistar Institute of Anatomy and Biology.

CHAPTER 4

Barlow, G. W. 1977. Modal action patterns. In *How Animals Communicate*, T. A. Sebeok, 98–134. Bloomington: Indiana University Press.

Coppinger, R., and L. Coppinger. 1996. Biological bases of behavior of domestic dog breeds. In *Readings in Companion Animal Behavior*, ed. V. Voith and P. Borchelt, 9–18. Trenton, NJ: Veterinary Learning Systems.

———. 1998. Differences in the behavior of dog breeds. In *Genetics and Behavior of Domestic Animals*, ed. T. Grandin. San Diego, CA: Academic Press.

———. 2001. *Dogs: A New Understanding of Canine Origin, Behavior and Evolution.* New York: Scribner.

Coppinger, R., and R. Schneider. 1995. The evolution of working dog behavior. In *The Domestic Dog*, ed. J. A. Serpell. Cambridge: Cambridge University Press.

Fentress, J. C., and P. J. McLeod. 1986. Motor patterns in development. In *Developmental Psychobiology and Developmental Neurobiology*, ed. E. M. Blass, 35–97. Handbook of Behavioral Neurobiology, vol. 8. New York: Plenum Press.

Lehner, P. 1979. *Handbook of Ethological Methods.* Cambridge: Cambridge University Press.

Leyhausen, P. 1979. *Cat Behavior: The Predatory and Social Behavior of Domestic and Wild Cats.* New York: Garland STPM Press.

Lord, K., R. Schneider, and R. Coppinger. In press. The evolution of working dog behavior. In *The Domestic Dog*, ed. J. A. Serpell. Cambridge: Cambridge University Press.

Morey, D. F. 1992. Size, shape, and development in the evolution of the domestic dog. *Journal of Archaeological Science* 19:181–204.

Schleidt, W. M. 1974. How "fixed" is the fixed action pattern? *Zeitschrift für Tierpsychologie* 36:184–211.

CHAPTER 5

Hall, W. G., and C. L. Williams. 1983. Suckling isn't feeding, or is it? a search for developmental continuities. *Advances in the Study of Behavior* 13:219–54.

Leyhausen, P. 1979. *Cat Behavior: The Predatory and Social Behavior of Domestic and Wild Cats.* New York: Garland STPM Press.

Lord, K. 2010. A heterochronic explanation for the behaviorally polymorphic genus

Canis: a study of the development of behavioral difference in dogs (*Canis lupus familiaris*) and wolves (*Canis lupus lupus*). PhD diss. University of Massachusetts, Amherst.

———. 2013. A comparison of the sensory development of wolves (*Canis lupus lupus*) and dogs (*Canis lupus familiaris*). *Ethology* 119:110–20.

Lord, K., M. Feinstein, B. Smith, and R. Coppinger. 2013. Variation in reproductive traits of members of the genus *Canis* with special attention to the domestic dog (*Canis familiaris*). *Behavioural Processes* 92:131–42.

CHAPTER 6

Berridge, K. C., J. C. Fentress, and H. Parr. 1987. Natural syntax rules control action sequence of rats. *Behavioural Brain Research* 23:59–68.

Coppinger, R., and L. Coppinger. 2001. *Dogs: A New Understanding of Canine Origin, Behavior and Evolution.* New York: Scribner.

Coppinger, R. P., C. K. Smith, and L. Miller. 1985. Observations on why mongrels may make effective livestock protecting dogs. *Journal of Range Management* 38:560–61.

Coren, S. 1995. *The Intelligence of Dogs: A Guide to the Thoughts, Emotions, and Inner Lives of Our Canine Companions.* New York: Bantam Books.

Fentress, J. C. 1990. Organizational patterns in action: local and global issues in action pattern formation. In *Signal and Sense: Local and Global Order in Perceptual Maps,* ed. G. M. Edelman, W. E. Gall, and W. M. Cowan, 357–82. New York: Wiley-Liss.

Fentress, J. C., and S. Gadbois. 2001. The development of action sequences. In *Developmental Psychobiology, Developmental Neurobiology and Behavioral Ecology: Mechanisms and Early Principles,* ed. E. M. Blass. Handbooks of Behavioral Neurobiology, vol. 13: New York: Kluwer Academic Publishers.

Gadbois, S., O. Sievert, C. Reeve, F. H. Harrington, and J. C. Fentress. 2015. Revisiting the concept of behavior patterns in animal behavior with an example from food-caching sequences in wolves (*Canis lupus*), coyotes (*Canis latrans*), and red foxes (*Vulpes vulpes*). *Behavioural Processes* 110:3–14.

Leyhausen, P. 1973. *Motivation of Humans and Animal Behavior: An Ethological View.* New York: Van Nostrand.

Lord, K., R. P. Coppinger, and L. Coppinger. 2013. Differences in the behavior of dog breeds. In *Genetics and the Behavior of Domestic Animals,* ed. T. Grandin and M. J. Deesing 195–235. 2nd ed. San Diego, CA: Academic Press.

Schleidt, W. M. 1974. How "fixed" is the fixed action pattern? *Zeitschrift für Tierpsychologie* 36:184–211.

Serpell, J., and J. A. Jagoe. 1995. Early experience and the development of behaviour. In *The Domestic Dog: Its Evolution, Behaviour, and Interactions with People,* ed. J. Serpell, 79–102. Cambridge: Cambridge University Press.

Twitty, V. C. 1966. *Of Scientists and Salamanders.* San Francisco: W. H. Freeman.

CHAPTER 7

Bateson, P. 1979. How do sensitive periods arise and what are they for? *Animal Behaviour* 27:470–86.

———. 2010. *Independent Inquiry into Dog Breeding.* Halesworth, UK: Micropress.

Chomsky, N. 1975. *Reflections on Language.* New York: Pantheon Press.

Estep, D. Q. 1996. The ontogeny of behavior. In *Readings in Companion Animal Behavior*, ed. V. L. Voith and P. L. Borchelt, 19–31. Trenton, NJ: Veterinary Learning Systems.

Fentress, J. C., and F. P. Stillwell. 1973. Grammar of a movement sequence in inbred mice. *Nature* 244:52–53.

Fox, M. 1969. Behavioral effects of rearing dogs with cats during the "critical period of socialization." *Behaviour* 35:273–80.

Goldin-Meadow, S. 2005. Watching language grow. *Proceedings of the National Academy of Science* 102:2271–72.

Lord, K. A. 2013. A comparison of the sensory development of wolves (*Canis lupus lupus*) and dogs (*Canis lupus familiaris*). *Ethology* 119:110–20.

Serpell, J., and J. A. Jagoe. 1995. Early experience and the development of behaviour. In *The Domestic Dog: Its Evolution, Behaviour, and Interactions with People*, ed. J. Serpell, 79–102. Cambridge: Cambridge University Press.

Schneider, R. A. 2007. How to tweak a beak: molecular techniques for studying the evolution of size and shape in Darwin's finches and other birds. *BioEssays* 29:1–6.

Scott, J. P., and M. Marston. 1950. Critical periods affecting the development of normal and maladjusted social behavior of puppies. *Journal of Genetic Psychology* 77:25–60.

Spencer, J. P., M. S. Blumberg, B. McMurray, S. R. Robinson, L. K. Samuelson, and J. B. Tomblin. 2009. Short arms and talking eggs: why we should no longer abide the nativist-empiricist debate. *Child Development Perspectives* 3, no. 2: 79–87

Twitty, V. C. 1966. *Of Scientists and Salamanders.* San Francisco: W. H. Freeman.

West, M, A. King, and D. White. 2003. The case for developmental ecology. *Animal Behaviour* 66:617–22.

West-Eberhard, M. J. 2003. *Developmental Plasticity and Evolution.* Oxford: Oxford University Press.

CHAPTER 8

Altenberg, L. 1994. Emergent phenomena in genetic programming. In *Evolutionary Programming: Proceedings of the Third Annual Conference*, ed. A. V. Sebald and L. J. Fogel, 233–41. River Edge, NJ: World Scientific Publishing.

Bell, H. S., and M. Pellis. 2011. A cybernetic perspective on food protection in rats: simple rules can generate complex and adaptable behaviour. *Animal Behaviour* 82:659–66.

Coppinger, R., and M. Feinstein. 1991. Hark! hark! the dogs do bark. and bark and bark. *Smithsonian* 21:119–29.

Escobedo, R., C. Muro, L. Spector, and R. P. Coppinger. 2014. Group size, individual role differentiation and effectiveness of cooperation in a homogeneous group of hunters. *Journal of the Royal Society: Interface* 11, no. 95: 1–10.

Fentress, J. C. 1992. Emergence of pattern in the development of mammalian movement sequences. *Journal of Neurobiology* 23:1529–56. doi: 10.1002/neu.480231011.

Lord, K., M., Feinstein, and R. Coppinger. 2009. Barking and mobbing. *Behavioural Processes* 81:358–68.

Morton, E. S. 1977. On the occurrence and significance of motivation-structural rules in some bird and mammal sounds. *American Naturalist* 111:855–69.

Muro, C., R. Escobedo, L. Spector, and R. P. Coppinger. 2011. Wolf-pack (*Canis lupus*) hunting strategies emerge from simple rules in computational simulations. *Behavioural Processes* 88:192–97.

Saunders, P. T. 1993. The organism as a dynamical system. *Thinking about Biology*, ed. W. Stein and F. Varela, 41–63. SFI Studies in the Sciences of Complexity, Lecture Notes, vol. 3. Reading, MA: Addison Wesley.

Schassburger, R. 1993. *Vocal Communication in the Timber Wolf, Canis lupus (Linnaeus): Structure, Motivation, and Ontogeny.* Advances in Ethology, no. 30. Berlin: Paul Parey.

Spector, L. 2011. Towards practical autoconstructive evolution: self-evolution of problem-solving genetic programming systems. In *Genetic Programming Theory and Practice, vol. 8*, ed. R. Riolo, T. McConaghy, and E. Vladislavleva, 17–33. New York: Springer.

Spector, L., J. Klein, C. Perry, and M. Feinstein. 2003. Emergence of collective behavior in evolving populations of flying agents. In *Proceedings of the Genetic and Evolutionary Computation Conference (GECCO-2003)*, ed. E. Cantu-Paz et al., 61–73. Berlin: Springer.

Strömböm D., R. P. Mann, A. M. Wilson, S. Hailes, A. J. Morton, D. J. T. Sumpter, and A. J. King. 2014. Solving the shepherding problem—heuristics for herding autonomous, interacting agents. *Journal of the Royal Society: Interface* 11:201407819.

Zimen, E. 1987. Ontogeny of approach and flight behavior toward humans in wolves, poodles and wolf-poodle hybrids. In *Man and Wolf*, ed. H. Frank, 275–92. Dordrecht: Dr. W. Junk Publishers.

CHAPTER 9

Bekoff, M. 1995. Play signals as punctuation: the structure of social play in canids. *Behaviour* 132:419–29.

Bekoff, M., and J. A. Byers. 1981. A critical re-analysis of the ontogeny and phylogeny of mammalian social and locomotor play: an ethological hornet's nest. In *Behavioral Development: The Bielefeld Interdisciplinary Project*, edited by K. Immelmann et al., 296–337. Cambridge: Cambridge University Press.

————, eds. 1998. *Animal Play: Evolutionary, Comparative, and Ecological Approaches*. New York: Cambridge University Press.

Bell, H. C., and S. M. Pellis. 2011. A cybernetic perspective on food protection in rats: simple rules can generate complex and adaptable behaviour. *Animal Behaviour* 82:4.

Bradshaw, J. W. S., A. J. Pullen, and J. Nicola. 2015. Why do adult dogs "play"? *Behavioural Processes* 110:82–87.

Burghardt, G. M. 1998. The evolutionary origins of play revisited: lessons from turtles. In *Animal Play: Evolutionary, Comparative, and Ecological Perspectives*, ed. M. Bekoff and J. A. Byers, 1–26. Cambridge: Cambridge University Press.

————. 2009. *The Genesis of Animal Play*. Cambridge, MA: MIT Press.

————. 2011. Defining and recognizing play. In *The Oxford Handbook of the Development of Play*, A. D. Pellegrini, 9–18. Oxford: Oxford University Press.

————. 2014. A brief glimpse at the long evolutionary history of play. *Animal Behavior and Cognition* 1:90–98.

————. In press. The origins, evolution, and interconnections of play and ritual: setting the stage. In *Play, Ritual and Belief an Animals and in Early Human Societies*, ed. C. Renfrew, I. Morley, and M. Boyd. Cambridge: Cambridge University Press.

Coppinger, R. P., J. Glendinning, E. Torop, C. Matthay, M. Sutherland, and C. Smith. 1987. Degree of behavioral neoteny differentiates canid polymorphs. *Ethology* 75:89–108.

Coppinger, R. P., and C. K. Smith. 1989. A model for understanding the evolution of mammalian behavior. *Current Mammalogy* 2:335–74.

Fagen, R. 1981. *Animal Play Behavior*. New York: Oxford University Press.

————. 1992. Play, fun and the communication of well-being. *Play and Culture* 5:40–58.

Fagen, R., and J. Fagen. 2004. Juvenile survival and benefits of play behaviour in brown bears, *Ursus arctos*. *Evolutionary Ecology Research* 6:89–102.

————. 2009. Play behaviour and multi-year juvenile survival in free-ranging brown bears, *Ursus arctos*. *Evolutionary Ecology Research* 11:1053–67.

Graham, K. L., and G. M. Burghardt. 2010. Current perspectives on the biological study of play: signs of progress. *Quarterly Review of Biology* 85:393–418.

Leyhausen, P., 1979. *Cat Behavior: The Predatory and Social Behavior of Domestic and Wild Cats*. New York: Garland STPM Press.

Palagi, E., G. M. Burghardt, B. Smuts, G. Cordoni, S. Dall'Olio, H. N. Fouts, M. Řeháková-Petrů, S. M. Siviy, and S. M. Pellis. 2015. Rough-and-tumble play as a window on animal communication. *Biological Reviews*.

Panksepp, J. 1981. The ontogeny of play in rats. *Developmental Psychobiology* 14, no. 4: 327–32.

————. 1998. *Affective Neuroscience: The Foundations of Human and Animal Emotions*. New York: Oxford University Press.

Pellis, S. M., and V. C. Pellis. 1996. On knowing it's only play: the role of play signals in play fighting. *Aggression and Violent Behavior* 1:249–68.

————. 2009. *The Playful Brain: Venturing to the Limits of Neuroscience.* Oxford: Oneworld Press.

Pellis, S. M., V. C. Pellis, and H. C. Bell. 2010. The function of play in the development of the social brain. *American Journal of Play* 2, no. 3: 278–96.

Pellis, S. M., V. C. Pellis, and C. J. Reinhart. 2010. The evolution of social play. In *Formative Experiences: The Interaction of Caregiving, Culture, and Developmental Psychobiology,* ed. C. Worthman, P. Plotsky, D. Schechter, and C. Cummings, 404–31. Cambridge: University Press, Cambridge.

Richmond, G., and B. D. Sachs. 1980. Grooming in Norway rats: the development and adult expression of a complex motor pattern. *Behaviour* 75, nos. 1–2: 82–96.

Spencer, H. 1855. *The Principles of Psychology.* London: Longman, Brown Green and Longmans.

CHAPTER 10

Arons, C., and W. Shoemaker. 1992. The distribution of catecholamines and β-endorphin in the brains of three behaviorally distinct breeds of dogs and their F_1 hybrids. *Brain Research* 594, no. 1: 31–39.

Baron-Cohen, S. 1991. Precursors to a theory of mind: understanding attention in others. In *Natural Theories of Mind: Evolution, Development and Simulation,* ed. A. Whiten, 233–51. Oxford: Blackwell.

Bekoff, M., C. Allen, and G. Burghardt. 2002. *The Cognitive Animal.* Cambridge, MA: MIT Press.

Bensky, M. K., S. D. Gosling, and D. L. Sinn. 2013. The world from a dog's view: a comprehensive review of dog cognition research. In *Advances in the Study of Behavior,* ed. H. J. Brockmann, 209–387. Vol. 45. Amsterdam: Elsevier.

Collier-Baker, E., J. M. Davis, and T. Suddendorf. 2004. Do dogs (*Canis familiaris*) understand invisible displacement? *Journal of Comparative Psychology* 118:421–33.

Dennett, D. 1991. *Consciousness Explained.* New York: Little Brown and Co.

Fisher, S. E. 2006. Tangled webs: tracing the connections between genes and cognition. *Cognition* 101:270–97.

Fiset, S., and V. Plourde. 2012. Object permanence in domestic dogs (*Canis lupus familiaris*) and gray wolves (*Canis lupus*). *Journal of Comparative Psychology* 127, no. 2: 115–27. doi: 10.1037/a0030595.

Gagnon, S., and F. Y. Doré. 1993. Search behavior of dogs (*Canis familiaris*) in invisible displacement problems. *Animal Learning and Behavior* 21:246–54.

Griffin, D. R. 1976. *The Question of Animal Awareness.* New York: Rockefeller University Press.

————. 1984. *Animal Thinking.* Cambridge, MA: Harvard University Press.

————. 1992. *Animal Minds.* Chicago: University of Chicago Press.

Hall, N. J., M. A. R. Udell, N. R. Dorey, A. L. Walsh, and C. D. L. Wynne. 2011. Megachiropteran bats (*Pteropus*) utilize human referential stimuli to locate hidden food. *Journal of Comparative Psychology* 125:341–46.

Hare, B., M. Brown, C. Williamson, and M. Tomasello. 2002. The domestication of social cognition in dogs. *Science* 298:1634–36.

Horowitz, A. 2009. Attention to attention in domestic dog (*Canis familiaris*) dyadic play. *Animal Cognition* 12:107–18.

————, ed. 2014. *Domestic Dog Cognition and Behavior*. Berlin: Springer.

Jackendoff, R. 1994. *Patterns in the Mind*. New York: Basic Books.

Kendrick, K. M. 1991. How the sheep's brain controls the visual recognition of animals and humans. *Journal of Animal Science* 69:5008–16.

Kruska D. 1988. Mammalian domestication and its effect on brain structure and behavior. In *Intelligence and Evolutionary Biology*, ed. H. J. Jerison and I. Jerison. New York: Springer-.

MacNeilage, P. F., 2011. Lashley's serial order problem and the acquisition/evolution of speech. *Cognitive Critique* 3:49–83.

McGinn, C. 2000. *The Mysterious Flame: Conscious Minds in a Material World*. New York: Basic Books.

Miklósi, Á. 2009. *Dog Behaviour, Evolution, and Cognition*. 1st ed. New York: Oxford University Press.

Miklósi, Á., R. Polgárdi, J. Topál, and V. Csányi. 2000. Intentional behaviour in dog-human communication: an experimental analysis of "showing" behaviour in the dog. *Animal Cognition* 3:159–66.

Miklósi, Á., and K. Soproni. 2006. A comparative analysis of animals' understanding of the human pointing gesture. *Animal Cognition* 9:81–93.

Purves, D. 1988. *Body and Brain*. Cambridge, MA: Harvard University Press.

Shettleworth, S. J. 1998. *Cognition, Evolution and Behavior*. New York: Oxford University Press.

Stillings, N., S. Weisler, C. Chase, M. Feinstein, J. Garfield, and E. Rissland. 1995. *Cognitive Science: An Introduction*. 2nd ed. Cambridge, MA: MIT Press.

Udell, M. A. R., N. R. Dorey, and C. D. L. Wynne. 2008. Wolves outperform dogs in following human social cues. *Animal Behaviour* 76:1767–73.

Udell, M. A. R., K. A. Lord, E. N. Feuerbacher, and C. D. L. Wynne. 2014. A dog's-eye view of canine cognition. In *Domestic Dog Cognition and Behavior*, ed. A. Horowitz, 221–40. doi: 10.1007/978-3-642-53994-7_10.

van Rooijen, J. 2010. Do dogs and bees possess a "theory of mind"? *Animal Behaviour* 79:e7–e8.

Wynne, C. D. L. 2004. *Do Animals Think?* Princeton, NJ: Princeton University Press.

Wynne, C. D. L., and M. A. R. Udell. 2014. *Animal Cognition: Evolution, Behavior and Cognition*. 2nd ed. New York: Palgrave Macmillan.

Wynne, C. D. L., M. A. R. Udell, and K. A. Lord. 2008. Ontogeny's impacts on human-dog communication. *Animal Behaviour* 76:e1–e4. doi:10.1016/j.anbehav.2008.03.010.

INDEX

Page numbers in *italics* indicate figures. Plates are designated as *Plate* and the number.

anthropomorphism: animal geniuses as belief in, 198–201; applied to machines, 187; applied to play behavior, 159; definition, 10; mythology and effects of, *Plate 1*, xi, 10–11, 12; in observing failure to RETRIEVE, 109–10; in observing wolves hunting, 136. *See also* popular mythologies

apes, 203

Aristotle, 137

Arons, Cynthia, 51

artificial selection, 50–51, 58, 211. *See also* breeds and breeding

assistance dogs. *See* service dogs

attentional capacity, 207–8

auditory system: capacity of, 103–4; human language acquisition reliant on, 127. *See also* LOST CALL

automata, 3–6, 10, 113. *See also* machines

Babbage, Charles, 189

"ball dog," 41–46, *42, 45*, 52

BARK and barking: complexity of, 136, 151–52; as emergent behavior, 157; frequency of, 67; mixed TONAL and NOISY motor pattern of, 153–57, *155, 156*; normal in deaf Border collies, 67–68, 127, 151; "scent bark," 153, *154*; as species-specific characteristic, 73–74. *See also* LOST CALL

Barlow, George W., 57

beagles, 89, 153, *154, 155*

beavers, 186

behavior: as adaptive, 2, 19–21, 106, 162; changes in shape result in changes in, 116–31; defined as shape moving in time and space, 2–5, 7–8, 41, 43, 55–57, 78, 116, 141, 208; genetic shaping of, 6–8; growth and play in relation to, 171–72, *173*, 174–81; hypertrophied type of, 151; key questions about, 33, 34–36; "species-typical" or "breed-typical," 53, 55;

summary, 211–15; as "taxonomic characteristic of species," 19, 56, 61, 65–66, 68, 141–42, 151, 195, 212. *See also* accommodative alterations and intrinsic rules interactions; emergence and emergent behavior; intrinsic motor patterns; motor patterns

Bekoff, Marc, 167

Belford, Charlie, 26

Bell, Heather, 149–50

Bergen, Bonnie, 128

biological systems: complexity of, 5–6, 51, 214–15; developmental changes in physiology, 171–72, *173*; genes and behavior of, 6–8; natural and artificial selection as, 50–51, 58, 211; surface area of animal in relation to, 44. *See also* brain and brain shape; sensory input systems; shape

birds: chickens, 75, 168, 170; courtship behaviors, 5; flight behavior, 35–36, 141–45, *142*; flocking as predator avoidance, 143, 213; human cues for, *Plate 8*; imprinting studies of, 19, 125–26; turkeys, *Plate 8*, 160

birthing motor pattern, *Plate 4*, 99–101, 106–7. *See also* neonates

bison: grazing patterns of, 26–27; wolves' cooperative hunting of, 148, *149*, 169

BITE, 165

body. *See* shape

Bondeson, Jan, 11

bone structure of pelvic girdle, 47–48. *See also* brain and brain shape

Border collies: activity needs, 58; breeding for specific motor patterns, 59, 60; Castro Laboreiros compared with, 92–93; collecting and study of, 28; congenitally deaf litter of, 67–68, 127, 151; cross-fostering experiments with, 94–95; disqualifying motor patterns of, 88–89, 91, 92; dopamine levels in, 51, 52; experiment with roosters and, 168, 169; EYE motor pattern, Plate 2, 78–79; EYE > STALK > CHASE motor pattern, Plate 3, 88, 110–11, 111, 204–5; foraging motor pattern, 87; herding behavior, 150–51; mechanical behavior, 211–12; motor patterns noted, 65, 67, 74; play behavior, 177–78, 180–81; puppies of, 29; RETRIEVE motor pattern, 109, 186; words seemingly understood by, 198

Bradshaw, J. W. S., 161

brain and brain shape: activities generating endorphins in, 163; as biological machine, 2–3, 187–89, 191–92; differences in shape, 51–52, 52; growth and differences in skulls, 172, 173, 181–82; human language organ in, 126–28; of institution- vs. family-raised children, 130–31; motor pattern in relation to, 74–75; neonatal behavior in relation to, 97–98; role of emergence in organization of, 141; visual development and, 122–25, 124. See also information-processing systems; mind

breeds and breeding: for adaptation to environment, 37, 46–51; cross-breed and -species suckling, 98; differences in, 29, 58, 59, 91–93, 93, 151, 177, 204, 211; increased sizes within, 44–45; for intrinsic threshold desired by trainer, 110–11; landrace guarding dogs as, 28; similar behavioral characteristics in particular, 53. See also motor patterns; shape; and specific breeds

bulldogs, 37, 59

Burghardt, Gordon M., 159

Byers, John, 182

caching behavior, 185–87

Canada geese, 141–45, 142

Canadian Centre for Wolf Research (Halifax), 186

canid species: birthing motor pattern, Plate 4, 99–101, 106–7; caching behavior, 185–87; facial information used by, 195; FOREFOOT-STAB motor pattern in some, 65–67, 66; mixed TONAL and NOISY motor pattern in captivity, 156–57; pair bonding in, 146–47; play bow in, 167–68, 168, 169, 170–71; "scent bark," 153, 154; sensory development in, 129; similar size at birth, 177; WOOF (alarm) motor pattern of, 152–53, 153, 154. See also dogs; dog-wolf comparisons; neonates; puppies (dog); puppies (wolf); wolves; and specific types

carnivores: characteristics of members, 77–78; foraging motor patterns, Plate 2, 78–83, 79. See also predators

Castro Laboreiro (dog breed), 91–93, 93, 94

Cat Behavior (Leyhausen), 61, 174

cat species: foraging motor patterns, Plate 2, 77–83; FOREFOOT-STAB motor pattern in some, 65; fresh meat as necessity for some, 77; play behavior, 165, 174; visual deprivation studies of kittens, 123. See also cheetah; pumas

cattle dogs, 58–59. See also droving dogs; guarding dogs

cells, growth rules, 137–38, 140–41

cemeteries, dogs in, Plate 1, 11, 12

CHASE: atypical or lack of development, 182–83; balance point between EYE > STALK and, 110–11, 111, 143; breed differences in, 204–5; of cattle and droving dogs, 58–59, 60; collecting data on, 69–70; descrip-

environment (*continued*)

shape and behavior in relation to, 42–43, 44, 52–53, 212–15; size of birds in relation to, 144; visual development dependent on, 122–25, 124

epigenesis, defined, 117–19

Escobedo, R., 148

Estrela Mountain dog, 152

ethogram: collecting data "on the ground," 68–71; components of foraging motor pattern in, 81–82; definition and construction of, 61–63; frequency of movement in, 67–68; of motor-pattern repertoire of same size/age puppies, 177–78; ontogenetic onsets and offsets in, 108–10; quality of movement in, 63–67; sequence of movement in, 68; training dogs and, 89–91; vocalizations noted, 153–54. *See also* motor patterns

ethology: approach and questions in, 1–6, 33–36, 135–36, 214–15; behavior as adaptive dictum in, 2, 19–21, 106, 162; collecting and studying dogs in, 25–33; etymology of, 19; first popular book on, x–xi; nature vs. nurture considered, 21–24; practice of, 17–19; species-specific genetic characteristics at heart of, 7–8; terminology of, 20–21. *See also* accommodative alterations and intrinsic rules interactions; cognitive ethology; emergence and emergent behavior; Hampshire College; intrinsic motor patterns; motor patterns

ethology techniques: 1/0 technique, 70; ad libitum sampling, 68–69; all-occurrence sampling, 69; focal animal sampling, 69, 70–71; instantaneous method, 70; measurements of behavior, 56–57; multiple observers in, 70–71; time-sampling techniques, 70–71; video recordings, 70. *See also* ethogram

"evo-devo" concept, 183

evolution: ethological perspective on, 2, 19–21, 106, 162; timing and character of development in, 183. *See also* adaptation; artificial selection; behavior; motor patterns; natural selection; shape

eyeball studies, 115

EYE > STALK: Border collie example of, 59; clapping and shape of Border collie in, 211–12; commonness of, *Plate 2*; development of, 121; evolutionarily conservative and homologous, 86–87; as intrinsic, 74; ORIENT in relation to, 88; "prey" object differences, 88–89; as species-specific characteristic, 89–90, 90; timing of exhibiting, 94–95; V flight pattern compared with, 142–43. *See also* EYE > STALK > CHASE

EYE > STALK > CHASE: balance point in, *Plate 3*, 110–11, 111, 143; brain feedback from sequence, 163; cat's capture of mouse in, 165; of cheetah, 61; as emergent behavior, 150–51; interruption in sequence, 168, 169, 170; puppies' play behavior incorporating, 178–79, 180–81; typical predatory sequence of, 78–82, 79. *See also* Border collies; CHASE; EYE > STALK

Fagen, Robert, 171

fear response: exposure and reduction of, 128, 129, 130, 193; motor pattern associated with, 65; physiology of, 195–96. *See also* hazard-avoidance behaviors

feeding: human feeding of dogs, 5, 8–9, 31, 77, 78, 86, 121; neonatal motor pattern in, 84, 86–87, 90, 97–99,

guarding dogs (*continued*)
tion in critical development period,
131–34, *133*; as working vs. companion dogs, 58. *See also* herding dogs;
and specific breeds

Haldane, J. B. S., 45–46
Hall, W. G., 86
Hampshire College (Amherst, MA):
accommodative alterations and intrinsic rules interactions studies at,
119–20, *120*; cooperative program
with ranchers, 87–88, 93, 164, 183;
ethology and cognitive science program, 27–30, 87–88, 91; Farm Center of, 100; LOST CALL recorded at,
102–3, *103*; Torop's motor patterns
study at, 91–93, *93*
Hank (sheep killer), 169, 170
Hans (horse), 199–200
Hare, Brian, 205, 206
Hawking, Stephen, 198
hazard-avoidance behaviors: as adaptations, 2, 57, 192–93; being social as,
125, 143; difficult to observe in companion dogs, 77; GAPE and, 63–64;
necessity of, 34; of neonates, 101,
171, 176; play behavior in relation to,
136, 159, 162, 165–67, 179, 181; of
predators, 31, 49; WOOF alarm call
in, 152–53, *153*, 154. *See also* fear response
HEAD-SHAKE, 67, 80, 82–83
hearing. *See* auditory system
herding dogs: CHASE mechanism in,
204–5; collecting and study of, 25,
30; "correct" breed-typical motor
patterns measured, 88–89; cross-
species fostering of, 95; foraging
motor patterns, 89–90, *90*; guarding dogs' differences from, 91–93,
93; as heelers or circular, heading,
60; play behavior, 177–78; variability

in motor patterns, 58–59, 61. *See also*
guarding dogs; *and specific breeds*
homologies, 65–66, 86–87
honeybees, 105
Horowitz, Alexandra, 207–8
horses, 37
hounds, 89, 90. *See also* greyhounds
Hubel, David, 122–23, 124, 130
humans: body changes throughout
life, 23–24; consciousness of, 195–
98; dogs' hearing compared with,
104; facial information for, 194–95;
hand motor patterns, 20, 55–56;
institution- vs. family-raised children, 130–31; intentionality of, 168;
intrinsic growth rules for, 118–19;
language acquisition in, 126–28,
196–97; marathon running, 47;
mental states and felt experience
of, 193; object understanding of,
201–2, 203; play behavior, 161–62,
165, 171; sexual motor patterns, 108;
smile or grin motor pattern, 64–65;
social cues from, *Plate 8*, 205–8;
WALK motor pattern, 72–73, *73*,
74. *See also* information-processing
systems
Hungary, pastoralists of Puszta grasslands, 32
hunting dogs, 58, 59, 89, 130, 160. *See*
also cooperative hunting
huskies, 26, 38, 46–47, 52. *See also* sled
dogs

Iditarod Trail Race (AK), 26, 38, 47, 48–
49. *See also* sled dogs
imprinting studies, 19, 125–26, 132
information-processing systems: adjacent searching in, 203–4; faces
in, 194–95; human cues for, *Plate*
8, 205–8; language for, 196–97;
mechanisms and processes in, 190–
94, *191*, 213–14; mind as, 188–89

82, 88, 91, 92, 180; LICK, 179;
MOUTH, 179; ORIENT, 61, 78,
80–81, 84–85, 88, 177; POINT, 58;
POUNCE, 80–81; SMILE, 64–65;
WALK, 74. See also BARK and bark-
ing; CHASE; EYE > STALK; EYE >
STALK > CHASE; SUCK and suck-
ling
mountain lions. See pumas
MOUTH, 179
Müller-Schwarze, Dietland, 164
Muro, C., 148
myths. See popular mythologies

Namibia, cheetah predation of cattle
in, 82
National Institute of Standards and
Technology, 189
National Scientific Research Center and
College, 75
natural selection: concept, 2, 50–51;
ethological perspective on, 2, 19–21,
106, 162; foraging motor patterns
and, 80
nature vs. nurture discourse, 21–24, 30,
93–95. See also environment; genetics
nematode worm, 105
neonates: behavioral metamorphosis
of, 175, 175–76, 179; feeding pattern,
84, 86–87, 90, 97–99, 98; shape, behavior, and changes in, 171–72, 173.
See also LOST CALL; SUCK and suckling
nest building, Plate 4, 99–100
New Guinea singing dog, 124
Nobel Prizes, noted, 18–19, 56, 122
NOISY and TONAL vocalizations (and
rules for which to use), 153–57, 154,
155, 156
nursing motor pattern. See neonates

olfaction. See smell
Olympics (Munich, 1972), 47

ontogeny: behavioral metamorphosis in, 174–80, 175; critical timing
in, 94–95; foraging motor pattern
linked to, 81–82; motor pattern onsets and offsets and, 108–10; phylogeny's interaction with, 113, 117–18,
212–15; puppies' foraging pattern
sets and development of, 83–87;
Tinbergen's question about, 35.
See also genetics; phylogeny
ORIENT, 61, 78, 80–81, 84–85, 88, 177
Origin of Species (Darwin), x
Osten, Wilhelm von, 199
Ovcharkas (Transcaucasus), 177–78
oxen, 37

packs, alpha animal in, 13
pandas, 77
Panksepp, Jaak, 163
pastoralists: of Abruzzi mountain
range, 30–31, 33; clothing of, 31,
32; predator deterrence system of,
27–28. See also guarding dogs
Peake, Kirsty, 108
Pellis, Sergio, 149–50, 182
pet dogs. See companion dogs
Pfungst, Oskar, 199–200
phenotypes: mind as part of, 190–92,
191; questions about, 23–24; as sum
of physical and behavioral traits, 22,
115; variations in, 117–19
phylogeny: foraging motor patterns in,
81; ontogeny's interaction with, 113,
117–18, 212–15; Tinbergen's question about, 35. See also evolution;
ontogeny
Piaget, Jean, 202
pigeons, 37
pigs, 37
play behavior: behavioral metamorphosis underlying, 174–80, 175;
complexity of, 136; discourse and
questions about, 162–67; as emer-

play behavior (*continued*)
gent behavior, 160, 170–71, 175, 176, 178, 180, 181, 183; injury from, 165; nature of, *Plates 6–7*, 160–62; play bow in, 167–68, *168*, *169*, 170–71; as "seemingly nonfunctional," 159, 162; shape and behavior changes in relation to, 171–72, *173*, 174–81; value of, 181–83

POINT, 58, 110

pointers, 89, *90*, 110

police dogs, 58, 59. *See also* service dogs

"popper" vs. skimmer, 47–48

popular mythologies: dogs as "Man's Best Friend," 8–9, 206; dogs as "wolves in our midst," 13–15. *See also* anthropomorphism

Portuguese Estrela Mountain dog, 152

POUNCE, 80–81

predators: control of, 26, 30, 43, 82, 100; foraging motor patterns, *Plate 2*, 77–83, *79*; FOREFOOT-STAB motor pattern, 65–67, *66*; play objects for, 161; puppy's LOST CALL as attracting, 104; rules in response to, 71; sequence of motor patterns in hunting, 68; wolves' eating of sheep, 31

primates: object permanence understood, 203; play behavior of, 171; SMILE or grin motor pattern in, 64–65. *See also* chimpanzees

prosopagnosia, 195

pugs, 121, *122*

pumas (mountain lions): behavior for escape from, 166; foraging motor patterns, *Plates 2–3*, 77, 80–81, 82; play behavior, *Plate 7*; targeted for extermination, 26

puppies (dog): barking of, 67; behavioral metamorphosis of, 174–80, *175*; bottle- vs. mother-nursed, 120–21; critical period for socialization in, 129; cross-fostering of, 29, 94–95,

132–34, *133*; food best for sled dogs, 118; foraging motor-pattern sets, 84–87, *85*; motor-pattern repertoire of same size/age, 177–78; physiological development, 83–84; shape and behavior changes in growth, 171–72, *173*, *174*, 176–81; shape and size of skulls, 172, *173*, 181–82; temperature differences felt by, 105; on their own after weaning, *Plate 5*; visual deprivation's effects on, 122–25, *124*. *See also* BARK and barking; neonates; play behavior

puppies (wolf): care for, after weaning, *Plate 5*; feeding of wild wolf, 49; food-begging behavior, 85, *85*–86, 147, 178, *179*; foraging motor-pattern sets of, 84–87, *85*; physiological development of, 83–84; shape of skulls, 172, *173*

quail, 119–20, *120*, 160

"quck," 120, *120*

Queensland Blue Heeler, 89

rabies epidemic, 9

rats, 86, 108, 149–50, 174

REGURGITATE, 85, *85*–86, 147, 178, *179*

reproductive activities: caring for puppies after weaning, *Plate 5*; changes in suckling environment, 120–21; communication system in, 101–2, *103*; gestation (average 63 days), 5, 100; hormonal state in mother, 107–8; nest building, *Plate 4*, 99–100; parental activities of dogs vs. wolves, 13–14; play behavior and, 165; RETRIEVE motor pattern, 106–10, 186; reward for, 163. *See also* Lina; LOST CALL; neonates; puppies (dog); puppies (wolf)

Republic of the Congo, rabies epidemic, 9

in hunting motor pattern, 68; shape of, 48; shape of skulls, 172, 173; socialization timing and process, 15; starvation of, 49; survival rate, 212; targeted for extermination, 26; WOOF motor pattern, 152–53, 153. *See also* dog-wolf comparisons; puppies (wolf)

"wolves in our midst" myth, 13–15

WOOF (alarm), 152–53, 153, 154

working dogs: breeding for specific motor patterns, 58, 59; Castro Laboreiros as herding vs. guarding, 91–93, 93, 94; foraging behavior patterns, 89–95, 90; humans' expectations of, 57–58; "landrace" populations of, 28; rationale for studying, 25–30; variations in, 211. *See also* guarding dogs; herding dogs; hunting dogs; service dogs; sled dogs

working draft animals, 37

World Health Organization, 9

worm, nematode, 105

Wynne, Clive, 205–6

Yellowstone National Park, 26–27, 83

Yerkes, Robert, x

Yorkshire terriers, 108